Prevention through Design

Structural Steel Design

Instructor's Manual

DEPARTMENT OF HEALTH AND HUMAN SERVICES
Centers for Disease Control and Prevention
National Institute for Occupational Safety and Health

Disclaimer

Mention of any company or product does not constitute endorsement by NIOSH. In addition, citations to Web sites external to NIOSH do not constitute NIOSH endorsement of the sponsoring organizations or their programs or products. Furthermore, NIOSH is not responsible for the content of these Web sites.

Ordering Information

This document is in the public domain and may be freely copied or reprinted. To receive NIOSH documents or other information about occupational safety and health topics, contact NIOSH at

Telephone: 1–800–CDC–INFO (1–800–232–4636)
TTY: 1–888–232–6348
Web site: www.cdc.gov/info

or visit the NIOSH Web site at www.cdc.gov/niosh

For a monthly update on news at NIOSH, subscribe to *NIOSH eNews* by visiting www.cdc.gov/niosh/eNews.

DHHS (NIOSH) Publication No. 2013–136

July 2013

SAFER • HEALTHIER • PEOPLE™

Please direct questions about these instructional materials to the National Institute for Occupational Safety and Health (NIOSH):

Telephone: (513) 533–8302
E-mail: preventionthroughdesign@cdc.gov

Foreword

A strategic goal of the Prevention through Design (PtD) Plan for the National Initiative is for designers, engineers, machinery and equipment manufacturers, health and safety (H&S) professionals, business leaders, and workers to understand the PtD concept. Further, they are to apply these skills and knowledge to the design and redesign of new and existing facilities, processes, equipment, tools, and this organization of work. In accordance with the PtD Plan, this module has been developed for use by educators to disseminate the PtD concept and practice within the undergraduate engineering curricula.

John Howard, M.D.
Director, National Institute for
    Occupational Safety and Health
Centers for Disease Control and Prevention

Contents

Acknowledgments

Authors:

T. Michael Toole, Ph.D., P.E.
Daniel Treppel
Stephen Van Nosdall

The authors thank the following for their reviews:

NIOSH Internal Reviewers

Pamela E. Heckel, Ph.D., P.E.
Donna S. Heidel, M.S., C.I.H.
Thomas J. Lentz, Ph.D., M.P.H., C.I.H.
Rick Niemeier, Ph.D.
Andrea Okun, Ph.D.
Paul Schulte, Ph.D.
Pietra Check, M.P.H.
John A. Decker, Ph.D.
Matt Gillen, M.S., C.I.H.
Roger Rosa, Ph.D.

Peer and Stakeholder Reviewers

Joe Fradella, Ph.D.
Matthew Marshall, Ph.D.
Gopal Menon, P.E.
James Platner, Ph.D.
Georgi Popov, Ph.D.
Deborah Young-Corbett, Ph.D., C.I.H., C.S.P., C.H.M.M.

Introduction

This Instructor's Manual is part of a broad-based multi-stakeholder initiative, Prevention through Design (PtD). This module has been developed for use by educators to disseminate the PtD concept and practice within the undergraduate engineering curricula. PtD anticipates and minimizes occupational safety and health hazards and risks[*] at the design phase of products,[†] considering workers through the entire life cycle, from the construction workers to the users, maintenance staff, and, finally, the demolition team. The engineering profession has long recognized the importance of preventing occupational safety and health problems by designing out hazards. Industry leaders want to reduce costs by preventing negative safety and health consequences of poor designs. Thus, owners, designers, and trade contractors all have an interest in the final design.

This manual is for one of four PtD education modules to increase awareness of construction hazards. The modules support undergraduate courses in civil and construction engineering. The four modules cover the following:

1. Reinforced concrete design
2. Mechanical-electrical systems
3. **Structural steel design**
4. Architectural design and construction.

This manual is specific to a PowerPoint slide deck related to Module 3, **Structural steel design**. It contains learning objectives, slide-by-slide lecture notes, case studies, test questions, and a list of citations. It is assumed that the users are experienced professors/lecturers in schools of engineering. As such, the manual does not provide specifics on how the materials should be presented. However, background insights are described for most of the slides for the instructor's consideration.

Numerous examples of inadequate design and catastrophic failures can be found on the Internet. If time permits, have the students seek, share, and analyze appropriate and inadequate designs. The PtD Web site is located at www.cdc.gov/niosh/topics/ptd. The National Institute for Occupational Safety and Health (NIOSH) Fatality Assessment and Control Evaluation (FACE) Reports can be found at www.cdc.gov/niosh/face/. Occupational Safety and Health Administration (OSHA) Fatal Facts are available at www.osha.gov/OshDoc/toc_FatalFacts.html.

---

[*] A "hazard" is anything with the potential to do harm. A "risk" is the likelihood of potential harm from that hazard being realized.

[†] The term *products* under the PtD umbrella pertains to structures, work premises, tools, manufacturing plants, equipment, machinery, substances, work methods, and systems of work.

Learning Objectives and Overview

Photo courtesy of Thinkstock

Structural Steel Design

EDUCATION MODULE

Developed by T. Michael Toole, Ph.D., PE
Daniel Treppel
Stephen Van Nosdall
Bucknell University

NOTES TO INSTRUCTORS

This education module contains a series of slides to present the Prevention through Design (PtD) concept, another set that summarizes structural steel design principles, and a third set that illustrates applications of the PtD concept to real-world scenarios. It contains specific examples of common workplace hazards related to construction and illustrates ways design can make a difference. There are several case studies to facilitate class discussions.

This education module is intended to facilitate incorporation of the PtD concept into your structural steel design course. You may wish to supplement the information presented in this module and may assign projects, class presentations, or homework as time permits. Sections may be presented independently of the whole. Presentation times are approximate, based on our presentation experience.

To activate the features embedded in some slides, please "enable content," make this a "trusted document," and view the slides in "slide show" mode. To show the presentation file in slideshow mode, press F5. Each slide is accompanied by speaker notes that you can read aloud while the slide is projected on the screen. The audience does not see the speaker notes. When you click on "Use Presenter View" on the Slide Show tab, your monitor displays the speaker notes but the projected image does not.

Thank you for using this module. To report problems or to make suggestions, please contact the National Institute for Occupational Safety and Health (NIOSH):

Telephone: (513) 533–8302
E-mail: preventionthroughdesign@cdc.gov

SOURCE
Photo courtesy of Thinkstock

 **Guide for Instructors**

Topic	Slide numbers	Approx. minutes
Introduction to Prevention through Design (PtD)	5–29	45
Design, Detailing, and Fabrication Process	30–36	10
Erection Process	37–41	10
Examples of Prevention through Design	42–77	50
Recap	78–79	5
References and Other Sources	80–86	—

NOTES

The first two slides of the presentation provide acknowledgments and general information. Learning objectives are delineated on Slide 3. Slide 4 contains the Overview. Slides 5 through 29 introduce the PtD concept and can be covered in approximately 45 minutes. Slides 30 through 36 introduce steel design, detailing, and fabrication. Slides 37 through 41 cover steel erection. Slides 42 through 77 contain examples of PtD theory applied to steel design. A summary is provided on slide 78.

Learning Objectives

- Explain the Prevention through Design (PtD) concept.

- List reasons why project owners may wish to incorporate PtD in their projects.

- Identify workplace hazards and risks associated with design decisions and recommend design alternatives to alleviate or lessen those risks.

NOTES

After completing this education module, you should be able to:

- Explain the PtD concept
- Describe motivations, barriers, and enablers for implementing PtD in projects
- List three reasons why PtD improves business value.

Overview

- PtD Concept

- Steel Design, Detailing, and Fabrication Process

- Steel Erection Process

- Specific Steel PtD Examples

Photo courtesy of Thinkstock

Structural Steel

NOTES

This is an overview of the PtD topics that we will cover. Many of you are probably not familiar with PtD, so we will spend a few minutes discussing what the concept is. Next I will summarize the structural design, detailing, and fabrication process. Similarly, I will summarize the steel erection process. Finally, I will show you some specific ways that structural engineers and detailers can incorporate PtD into their steel designs.

SOURCE

Photo courtesy of Thinkstock

Introduction to
Prevention through Design (PtD)

Introduction to Prevention through Design
EDUCATION MODULE

NOTES

Let's start by introducing PtD.

Occupational Safety and Health

- Occupational Safety and Health Administration (OSHA)
  www.osha.gov
  - Part of the Department of Labor
  - Assures safe and healthful workplaces
  - Sets and enforces standards
  - Provides training, outreach, education, and assistance
  - State regulations possibly more stringent

- National Institute for Occupational Safety and
  Health (NIOSH) www.cdc.gov/niosh
  - Part of the Department of Health and Human Services, Centers
    for Disease Control and Prevention
  - Conducts research and makes recommendations for the
    prevention of work-related injury and illness

Structural Steel

NOTES

All employers, including structural design firms, are required by law to provide their employees with a safe work environment and training to recognize hazards that may be present. They also must provide equipment or other means to minimize or manage the hazards.

Designers historically have not been familiar with the federal Occupational Safety and Health Act (OSH Act) standards because they were rarely exposed to construction jobsite hazards. However, with the increasing roles that designers are playing on worksites, such as being part of a design-build team, it is becoming increasingly important that they receive construction safety training, including information about federal and state construction safety standards.

The Occupational Safety & Health Act of 1970, Public Law 91-596 (OSH Act) [29 USC* 1900], was passed on December 29, 1970, "To assure safe and healthful working conditions for working men and women; by authorizing enforcement of the standards developed under the Act; by assisting and encouraging the States in their efforts to assure safe and healthful working conditions; by providing for research, information, education, and training in the field of occupational safety and health; and for other purposes." The construction industry standards

---

*United States Code. See USC in Sources.

enforced by the Occupational Safety and Health Administration (OSHA) are found in Title 29 Part 1926 of the Code of Federal Regulations [29 CFR 1926].

The National Institute for Occupational Safety and Health (NIOSH) is part of the Department of Health and Human Services, Centers for Disease Control and Prevention. The National Occupational Research Agenda (NORA) is a partnership program to stimulate innovative research and improved workplace practices. Unveiled in 1996, NORA has become a research framework for NIOSH and the nation. Diverse parties collaborate to identify the most critical issues in workplace safety and health. Partners, then, work together to develop goals and objectives for addressing these needs. Participation in NORA is broad, including stakeholders from universities, large and small businesses, professional societies, government agencies, and worker organizations. NIOSH and its partners have formed ten NORA Sector Councils: Agriculture, Forestry & Fishing; Construction; Healthcare & Social Assistance; Manufacturing; Mining; Oil and Gas Extraction; Public Safety; Other Services; Transportation, Warehousing & Utilities; and Wholesale and Retail Trade. The mission of the NIOSH research program for the Construction sector is to eliminate occupational diseases, injuries, and fatalities among individuals working in these industries through a focused program of research and prevention.

SOURCES

CFR. Code of Federal Regulations. Washington, DC: U.S. Government Printing Office, Office of the Federal Register.

NIOSH FACE reports [www.cdc.gov/niosh/face/]

OSHA Fatal Facts [www.setonresourcecenter.com/MSDS_Hazcom/FatalFacts/index.htm]

OSHA home page [www.osha.gov/pls/oshaweb/owastand.display_standard_group?p_toc_level=1&p_part_number=1926]

USC. United States Code. Washington, DC: U.S. Government Printing Office.

Construction Hazards

 **Construction Hazards**

- Cuts

- Electrocution

- Falls

- Falling objects

- Heat/cold stress

- Musculoskeletal disease

- Tripping

[BLS 2006; Lipscomb et al. 2006]

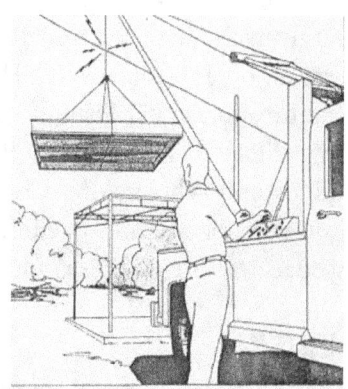

Graphic courtesy of OSHA

Structural Steel

NOTES

A construction worksite by its nature involves numerous potential hazards. A portion of the work is directly affected by weather. Workers interact with heavy equipment and materials at elevated heights, in below-ground excavations, and in multiple awkward positions. The composition of the site workforce changes over the project, and work is done autonomously at times and in coordination at others. The construction worksite is unforgiving to poor planning and operational errors.

For these reasons, pre-job construction-phase planning is used as a best practice to systematically address potential hazards. Project-specific worker safety orientations prior to site work also play an important role. PtD practices, by systematically looking further upstream at design-related potential hazards, extend these pre-job measures. PtD can help identify potential hazards so that they can be eliminated, reduced, or communicated to contractors for pre-job planning.

Every hazard that can be addressed should be addressed. Falling can cause serious injury. Boilermakers, pipe-fitters, and iron workers can experience career-ending musculoskeletal injuries by lifting heavy loads or working in a cramped position. Anyone can be seriously injured by a falling object. Whether a structural member or a simple wrench, a falling object can be deadly. Anyone can trip, but the elevated height and proximity to dangerous equipment increase the risk of injury on a construction site. Some accidents are caused by poor lighting and/or sunlight glare. Common injuries due to spatial misperception include hitting your head or cutting yourself on sharp corners. Hot summer and cold winter days can affect worker health. Personal protective equipment (PPE), such as hardhats, gloves, ear protection, and safety glasses, is required for a reason! Not every hazard on a construction worksite can be "designed out," but many significant ones can be minimized during the design phase.

SOURCES

BLS [2006]. Injuries, illnesses, and fatalities in construction, 2004. By Meyer SW, Pegula SM. Washington, DC: U.S. Department of Labor, Bureau of Labor Statistics, Office of Safety, Health, and Working Conditions.

Lipscomb HJ, Glazner JE, Bondy J, Guarini K, Lezotte D [2006]. Injuries from slips and trips in construction. Appl Ergonomics 37(3):267–274.

OSHA [ND]. Fatal facts accident reports index [foreman electrocuted]. Accident summary no. 17 [www.setonresourcecenter.com/MSDS_Hazcom/FatalFacts/Index.htm].

Graphic Courtesy of OSHA

ACCIDENT REPORT

ACCIDENT SUMMARY No. 17

Accident Type:	Electrocution
Weather Conditions:	Sunny, Clear
Type of Operation:	Steel Erection
Size of Work Crew:	3
Collective Bargaining	No
Competent Safety Monitor on Site:	Yes - Victim
Safety and Health Program in Effect:	No
Was the Worksite Inspected Regularly:	Yes
Training and Education Provided:	No
Employee Job Title:	Steel Erector Foreman
Age & Sex:	43-Male
Experience at this Type of Work:	4 months
Time on Project:	4 Hours

BRIEF DESCRIPTION OF ACCIDENT

Employees were moving a steel canopy structure using a "boom crane" truck. The boom cable made contact with a 7200 volt electrical power distribution line electrocuting the operator of the crane; he was the foreman at the site.

INSPECTION RESULTS

As a result of its investigation. OSHA issued citations for four serious violations of its construction standards dealing with training, protective equipment, and working too close to power lines.

OSHA's construction safety standards include several requirements which, If they had been followed here. might have prevented this fatality.

ACCIDENT PREVENTION RECOMMENDATIONS

1. Develop and maintain a safety and health program to provide guidance for safe operations (29 CFR 1926.20(b)(1)).
2. Instruct each employee on how to recognize and avoid unsafe conditions which apply to the work and work areas (29 CFR 1926.21(b)(2))
3. If high voltage lines are not de-energized, visibly grounded, or protected by insulating barriers, equipment operators must maintain a minimum distance of 10 feet between their equipment and the electrical distribution or transmission lines (29 CFR 1926.550(a)(15)(i)).

SOURCES OF HELP

*   Ground Fault Protection on Construction Sites (OSHA 3007) which describes OSHA requirements for electrical safety at construction sites.

- Construction Safety and Health Standards (OSHA 2207) which contains all OSHA job safety and health rules and regulations (1926 and 1910) covering construction
- OSHA Safety and Health Training Guidelines for Construction (available from the National Technical Information Service - Order No PB-239312/AS) comprised of a set of 15 guidelines to help construction employees establish a training program in the safe use of equipment, tools, and machinery on the job
- OSHA-funded free onsite consultation services Consult your telephone directory for the number of your local OSHA area or regional office for further assistance and advice (listed under the US Labor Department or under the state government section where states administer their own OSH programs).

NOTE: The case here described was selected as being representative of fatalities caused by improper work practices. No special emphasis or priority is implied nor is the case necessarily a recent occurrence. The legal aspects of the incident have been resolved, and the case is now closed.

Construction Accidents

 **Construction Accidents in the United States**

Construction is one of the most hazardous occupations. This industry accounts for

- 8% of the U.S. workforce, but 20% of fatalities

- About 1,100 deaths annually

- About 170,000 serious injuries annually

[CPWR 2008]

Photo courtesy of Thinkstock

Structural Steel

NOTES

As many of us know, construction is one of the most dangerous industries for workers. In the United States, construction typically accounts for 170,000 serious injuries and 1,100 deaths each year. The fatality rate is disproportionally high for the size of the construction workforce. Twenty percent of all collapses during construction are the result of structural design errors. Statistics like these do not tell the whole story. Behind every serious injury, there is a real story of an individual who suffered serious pain and may never fully recover. Behind every fatality, there are spouses, children, and parents who grieve every day for their loss. We all recognize that safety is a vital component of an inherently dangerous business. All of us—including architects and engineers—must do what we can to reduce the risk of injuries on our projects.

SOURCES

CPWR [2008]. The construction chart book. 4th ed. Silver Spring, MD: Center for Construction Research and Training.

New York State Department of Health [2007]. A plumber dies after the collapse of a trench wall. Case report 07NY033 [www.cdc.gov/niosh/face/pdfs/07NY033.pdf].

OSHA [ND]. Fatal facts accident reports index [laborer struck by falling wall]. Accident summary no. 59 [www.setonresourcecenter.com/MSDS_Hazcom/FatalFacts/Index.htm].

Photo courtesy of Thinkstock

NEW YORK

state department of

HEALTH

A Plumber Dies After the
Collapse of a Trench Wall
Case Report: 07NY033

SUMMARY

In May 2007, a 46 year old self-employed plumbing contractor (the victim) died when the unprotected trench he was working in collapsed. The victim was an independent plumber subcontracted to install a sewer line connection to the sewer main, part of a general contractor project to install a new sanitary sewer for an existing single family residence.

At approximately 12:30 PM on the day of the incident, the workers on site observed the victim walking back toward the residence for parts as they initiated their lunch break. When the victim did not come for his lunch or answer his cell phone, the general contractor and workers starting searching for the victim. The excavation contractor observed that a portion of the trench had collapsed where the victim was installing a sewer tap. The victim was found trapped in the trench under a large slab of asphalt, rock and soil. Three workers immediately climbed down the side of the trench to try to assist the victim. One of the workers called 911 on his cell phone. Police and emergency medical services (EMS) arrived on site within minutes. The EMS members entered the unprotected trench but could not revive the victim. The county trench rescue team recovered the victim's body at approximately seven feet below grade and lifted him from the ditch four hours after the incident. He was pronounced dead at the site. More than 50 rescue workers were involved in the recovery.

New York State Fatality Assessment and Control Evaluation (NY FACE) investigators concluded that, to help prevent similar occurrences, employers and independent contractors should:
- **Require that all employees, subcontractors, and site workers working in trenches five feet or more in depth are protected from cave-ins by an adequate protection system.**
- **Require that a competent person conducts daily inspections of the excavations, adjacent areas, and protective systems and takes appropriate measures necessary to protect workers.**
- **Require that all employees and subcontractors have been properly trained in the recognition of the hazards associated with excavation and trenching. In addition, the general contractor (GC) should be responsible for the collection and review of training records and require that all workers employed on the site have received the requisite training to meet all applicable standards and regulations for the scope of work being performed.**
- **Require that on a multi-employer work site, the GC should be responsible for the coordination of all high hazard work activities such as excavation and trenching.**

- **Require that all employees are protected from exposure to electrical hazards in a trench.**

Additionally,

- **Employers of law enforcement and EMS personnel should develop trench rescue procedures and should require that their employees are trained to understand that they are not to enter an unprotected trench during an emergency rescue operation.**
- **Local governing bodies and codes enforcement officers should receive additional training to upgrade their knowledge and awareness of high hazard work, including excavation and trenching. This skills upgrade should be provided to both new and existing codes enforcement officers.**
- **Local governing bodies and codes enforcement officers should consider requiring building permit applicants to certify that they will follow written excavation and trenching plans in accordance with applicable standards and regulations, for any projects involving excavation and trenching work, before the building permits can be approved.**

INTRODUCTION

In May, 2007, a 46 year old self-employed plumbing contractor died when the trench he was working in collapsed at a residential construction site. Approximately 8000 pounds of broken asphalt, rock and dirt fell atop the victim, fatally crushing him as he was installing a sewer tap to a town sewer main. The New York State Fatality Assessment and Control Evaluation (NY FACE) program learned about the incident from a newspaper article the following day. The Occupational Safety and Health Administration (OSHA) investigated the incident along with the county sheriff's office. The NY FACE staff met and reviewed the case information with the OSHA compliance officer. This report was developed based upon the information provided by OSHA, the county sheriff's department, and the county coroner's medical and toxicological reports.

The general contractor (GC) on the residential construction site had been hired by the homeowners to complete a project that included the installation of a new sanitary sewer connection for an existing single family residence. The GC was the owner and sole employee of his company, which had been in business for many years. The GC directed the work of two subcontractors on the work site to complete the installation of the residential sewer line.

- One subcontractor was an excavating company that had been in business for approximately four years. The owner of this company hired two workers to assist him with the excavation of the trench.
- The second subcontractor was the victim, a self-employed licensed plumber who had over twenty years of experience with a variety of construction projects, including the installation of sewer lines. The victim did not have any previous work relationship with either the GC or the excavation subcontractor.

The OSHA investigation report indicated that the GC and the subcontractor did not have health and safety programs. A formal health and safety plan had not been established to identify the hazards of the excavation project and the actions to be taken to remediate them. The GC, subcontractors and the subcontractors' employees did not have hazard recognition training or safety training on the fundamentals of excavation and trenching. None of the workers on the site were knowledgeable on excavation and trenching safety standards and applicable regulations and they did not understand the hazards and dangers associated with working in a trench. A competent person was not present to conduct initial and ongoing inspections of the excavation project, identify potential health and safety hazards such as possible cave-in, and oversee the use of adequate protection systems and work practices.

INVESTIGATION

The GC was hired to replace a crushed sewer line that ran under the driveway of an existing single family residence. Rather than dig up the driveway to replace the old line, which was thought to be more costly and time-consuming, the GC decided to run a new line. All required town permits had been obtained and the local codes enforcement requirements for one-call system notification of the excavation and underground utility location mark-outs had been completed. The work had been scheduled to be completed in one day (Friday), but the excavation subcontractor lost time due to hitting a water line and encountering very rocky soil during the excavation. The project had to be extended to two days (Friday and Monday). The town water and sewer inspector visited the work site on Friday, observed the digging of the trench which began at the residence, and halted the digging of the trench at the edge of the property to avoid having an open trench in the road and consequent road closure over a weekend. Excavation company workers had been observed in the trench spotting and hand digging.

On Monday, the day of the incident, the excavating subcontractor initiated excavation from the edge of the road to the sewer main in the roadway. An employee witness of the excavating company stated that the victim was directing excavation work while in the trench and hand digging to expose the sewer main once the excavator came close to the location. OSHA findings indicated that tools were uncovered in the trench in the area of the trench wall collapse, including a shovel, pick ax, hammer drill and drill bits, consistent with the scenario of the victim being in the ditch, hand digging to locate the sewer main. The town water and sewer inspector also visited the work site on Monday. He determined that the victim did not have the correct parts to complete the sewer connection, advised him of the correct parts, and indicated that he would return later in the day to re-inspect and photograph the completed sewer tap in order to allow the excavating subcontractor to run the pipe back to the house, backfill the excavation and reopen the road.

The GC left the work site to purchase the correct parts, while the excavation continued. The dimensions of the final trench were approximately 55 feet in length, 3 feet to 8 feet in depth, and 30 inches to 128 inches in width (see Figure 1). It was shaped like a "T." The gravity sewer main that the victim was connecting to was located at a depth of 7 feet 4 inches (7' 4") below grade at the east (E) end of the top of the "T." Installation of new sewer pipe from the residence had been initiated and some of the trench had already been backfilled. The length of the trench from the top of the "T" to the location of the newly installed sewer pipe was 35 feet 11 inches (35'11") at the time of the incident. Soil analysis results, conducted after the incident, indicated a granular, sandy gravel Type C soil (OSHA Excavation Standard) that contained large cobbles and boulders, the least stable soil type.

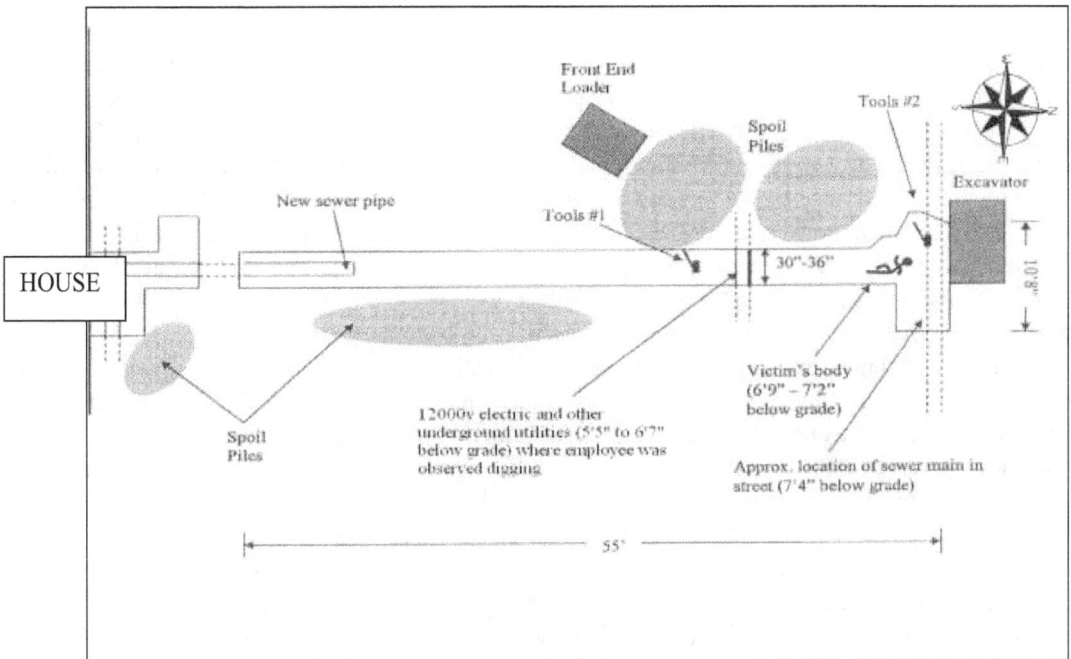

Figure 1: Schematic of the excavation and the incident site (courtesy of OSHA)

The faces of the trench were vertical. No shoring or benching was used. Large cobbles and boulders and loose rock/dirt were visible on the face of the excavation and were not removed or supported. The pavement above the E and W faces of the excavation had been undermined during excavation activities and no support system was utilized to protect employees from a possible collapse. Pieces of road pavement and asphalt had been undermined during excavation activities in the road in the proximity of the sewer main at the top of the "T." These areas were in plain view and did not have additional support. On the W side of the excavation, loose boulders, rock and debris in spoils piles were located less than two feet from the edge of the trench. (Figure 2) The excavator was positioned adjacent to the N end of the trench, where undermined areas were in plain sight. The N end of the trench, where the victim was installing the sewer tap, also lacked an access ladder or other safe means of entry/egress.

Figure 2: View of the west wall of the excavation south of the "T."
Note the boulders and loose rock/dirt on the excavation face as well as the location of the spoils pile within 2 feet of the edge of the trench. (courtesy of OSHA)

The GC returned just before 12 noon with the correct parts and handed them to the victim. The GC left the site in order to purchase lunch for the workers, including the victim. At this same time, the victim called the town water and sewer inspector, informed him that he had located the sewer main, had all the correct parts, and was ready to connect. The town inspector informed the victim that someone from the town would be out after lunch to inspect and photograph the sewer tap. According to the town inspector, a sewer tap to a sewer main is a simple job that would take about 20 minutes to complete. The GC returned with lunch at 12:30PM. The workers, with the exception of the victim, took a break for lunch at a location near the front end loader (Figure 1). The workers saw the victim walking in the trench in the direction of the residence and heard him say that he was "looking for a splitter for a three-way." By 1:00 PM the victim still had not come for his lunch. The GC called the victim on his cell phone and looked for him in his van behind the house. The other workers joined in the search. The excavating subcontractor observed that a portion of the west side of the trench had collapsed. When the workers approached the excavation, they found the victim trapped in the trench under a large slab of asphalt, rock, and soil, with only the back of his head exposed. Three workers climbed down the side of the trench to try to assist the victim.

The workers removed the dirt from around his head, lifted his head, and tried to clear his airway. They checked for a pulse, but found none. One of the workers then called 911 from his cell phone. The workers attempted to move the slab of asphalt without success. Within minutes, the police arrived, followed by EMS at approximately 1:08 PM. The EMS personnel entered the unprotected trench but were unable to revive the victim. Volunteer firefighters from multiple fire departments and a special trench rescue team responded, the latter team having been created by the county after the deaths of two workers in a construction trench collapse 10 years earlier. A wooden safety box was built by the trench rescue team and efforts began to free the victim from entrapment by chipping the asphalt slab into pieces. Using a system of ropes and pulleys, the rescue team lifted the victim from the ditch at 4:25 PM. His body had been recovered at about 7' below grade. The county coroner pronounced him dead at 4:35 PM. Approximately 50 rescuers responded to the 911 call.

The OSHA investigation resulted in findings that the trench section that collapsed was a triangular shaped area at the northwest corner of the excavation, approximately 5 feet 1 inch (5' 1") in length, 4 feet (4') wide, and 6-7 feet (6-7') deep. Multiple hazards were present, but had not been identified and remediated. The W side of the excavation collapsed and pieces of asphalt paving and rock fatally crushed the victim while he was making the sewer tap (Figures 3 and 4).

The hazards of the unprotected trench exposed additional people to the excavation collapse as the GC, the excavation company workers and EMS personnel entered the trench to attempt a rescue of the victim. In addition to the trench hazards, no precautions had been taken to prevent exposure to the underground electrical and utility lines. The town inspector had noted that a young employee of the excavation company was "manually hand digging with shovel and pick ax "within a few inches of the buried electrical lines." This is consistent with OSHA findings that indicated attempts had been made to cut the rock in the face of the trench at the location of the underground utilities. A demo saw, hammer drill and cordless reciprocating saw used to cut rocks and pavement were found within inches of the 12,000 volt underground electrical line. Several other utilities were also exposed in this location at the edge of the road (Figure #1, Tools #1). EMS personnel also entered the trench when power was still connected to the utilities in the trench.

Figure 3: Location of collapse.
Note spoils piles and equipment located less
2 feet from the edge of the trench
(courtesy of OSHA)

Figure 4: Area of trench collapse
Note the large boulders hanging from the than
excavation faces and undermined areas on the
edge of the trench (courtesy of OSHA)

RECOMMENDATIONS/DISCUSSION

Recommendation #1: *Employers and independent contractors should require that all employees, subcontractors and site workers working in trenches five feet or more in depth are protected from cave-ins by an adequate protection system.*

Discussion: Employers and contractors should require that all employees working in trenches five feet deep or more are protected from cave-ins by an adequate protection system appropriate to the conditions of the trench, including sloping techniques or support systems such as shoring or trench boxes (OSHA 29CFR 1926.652). Sloping involves positioning the soil away from an excavation trench at an angle that would prevent the soil from caving into the trench. Even in shallow trenches less than five feet in depth, the possibility of accidents still exists. Trenches five feet deep or less should also be protected if a competent person identifies a cave-in potential. Trench protection systems are available to all employers and independent contractors, even as rental equipment. Employers should also require that all pieces of excavated pavement, asphalt, dirt, rock, boulders, and debris as well as excavation equipment are located in spoils piles or positions that are at least two feet from the edge of the excavated trench. Where a two foot setback is not possible, spoils may need to be hauled to another location. In this incident, sloping would not have been an appropriate protection system, due to the composition of the soil. Employers and contractors should consult tables located in the appendices of the OSHA Excavation Standard that detail the protection required based upon the soil type and environmental conditions present at a work site. Employers and contractors can also consult with manufacturers of protective systems to obtain detailed guidance for the appropriate use of protection systems.

Trenches should be kept open only for the minimum amount of time needed. Hinze and Bren (1997) observed that the risk of a collapse in an unprotected trench increases the longer a trench is open. They propose that after a trench is dug, the apparent cohesion of trench walls may begin to relax after only four hours, contributing to increasingly unstable walls in an unprotected trench. In this incident, a 45 feet length of the trench had been excavated and was left open for more than two days. The trench section where the incident occurred was dug at approximately 8:30 AM on the day of the incident. Hand digging and incorrect parts resulted in additional delays in making the sewer tap to the main. The trench collapse occurred approximately four hours later, between 12:30 PM and 1:00 PM.

The key to preventing a trench accident is not to enter an unprotected trench. When the walls of a trench collapse or cave in, the results are entrapment or struck-by incidents to anyone caught inside, accidents which can occur in seconds. Many workers in a trench are in a kneeling or squatting position that results in little opportunity for an escape. Victims do not need to be completely covered in soil. Even with partial covering, enough pressure is created for mechanical asphyxia in which the weight of the dirt and soil compresses the chest. One cubic yard of soil has an average weight of 2500 pounds (Figure 4), but can vary due to the composition and moisture content.

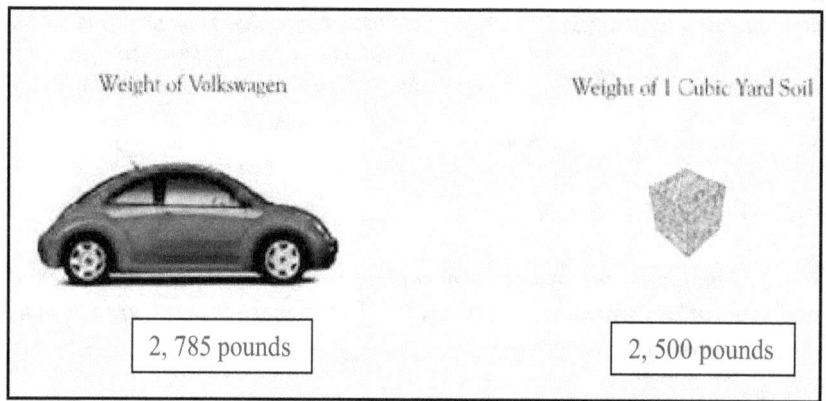

Figure 5: Weight of one cubic yard of soil (courtesy of "Weights of Building Materials, Agricultural Commodities, and Floor Loads for Buildings" standard reference)

Recommendation #2: *Employers and independent contractors should require that a competent person conducts daily inspections of the excavations, adjacent areas, and protective systems and takes appropriate measures necessary to protect workers.*

Discussion: Employers and independent contractors are responsible for complying with the OSHA Excavation Standard requirements to designate a competent person on site for excavation and trenching projects to make daily inspections of excavations, the adjacent areas, and protective systems (OSHA 29CFR 1926.651). A competent person is defined as someone who is capable of identifying existing and predictable hazards in the surroundings and working conditions that are dangerous to employees and who has the authorization to take prompt corrective measures to eliminate them. They should inspect the trenches daily, as needed throughout the work shift, and as conditions change (for example, heavy rainfall or increased traffic vibrations). These inspections should be conducted before worker entry, to ensure that there is no evidence of a possible cave-in, failure of a protective system, hazardous conditions such as spoils piles or equipment location, or hazardous atmosphere.

In particular, competent persons are required by OSHA to complete a competent person training curriculum, which could be an OSHA training program or an equivalent safety or trade organization training. The competent person needs be knowledgeable on the hazards associated with excavation and trenching, as well as the causes of injuries and the safe work practices and specific protective actions needed. Competent persons must also be experienced in excavation and trenching with a minimum of hands-on training in a demonstration trench or in a field component. The competent person needs to know the key points of the OSHA Excavation Standard, including the excavation standards and appendices, checklists, soils analysis and the components of a daily trenching inspection.

Having a competent person is a particularly acute problem among contracting companies that employ fewer than 10 workers. Of the National Institute for Occupational Safety and Health (NIOSH) FACE cases related to excavation and trenching, 88% were non-union companies with less than 10 workers. These small companies are not members of trade associations and are the least likely to employ trench safety protections and to have an adequately trained competent person or an excavation crew.

In this incident, no competent person was hired by the GC to conduct initial and ongoing inspections of the trench. The GC, excavating contractor, and excavation company employees did not possess an understanding of the hazards associated with excavation and trenching operations or a knowledge of the requirements of the OSHA Excavation Standard. No one on-site was qualified to function as the competent person.

Recommendation #3: *Employers and independent contractors should require that all employees and subcontractors have been properly trained in the recognition of the hazards associated with excavation and trenching. On a multi-employer work site, the GC should be responsible for the collection and review of training records and require that all workers employed on the site have received the requisite training to meet all applicable standards and regulations for the scope of work being performed.*

Discussion: Excavation and trenching is one of the most hazardous construction operations. Even with a competent person on site, workers in excavation and trenching operations are also in need of health and safety training, including basic hazard recognition and prevention. Workers should be able to identify the specific hazards associated with excavation and trenching, the reasons for using protective equipment and how to work in a trench safely. Workers should be trained not to enter an unprotected trench, even in a rescue attempt, since they place themselves at risk of becoming injured or killed. If necessary, projects should be delayed until training requirements are met and training records are provided.

In this case, the general contractor, excavation subcontractor, and excavation company employees did not demonstrate adequate knowledge of safe work practices in excavation and trenching. The limited training in proper excavation technique as well as inadequate hazard recognition and prevention training were critical to the failure to properly assess the hazards present and protect the trench.

Recommendation #4: *Employers and independent contractors should require that on a multi-employer work site, the GC should be responsible for the coordination of all high hazard work activities such as excavation and trenching.*

Discussion: The GC is responsible and accountable for the safety of all employees, subcontractors and workers on the site. Health and safety plans should be in place to formally address the hazards that may be encountered, including written plans to manage these hazards and protect the safety of all workers on the site.

In this incident, the GC did coordinate the work activities of the subcontractors and workers on the job, but health and safety plans were not addressed. The management of excavation and trenching hazards was left to a subcontractor who was not a competent person, knowledgeable or trained in the requirements of the OSHA Excavation Standard.

Recommendation #5: *Employers of law enforcement and EMS personnel should develop trench rescue procedures and should require that their employees are trained to understand that they are not to enter an unprotected trench during an emergency rescue operation.*

Discussion: Employers of law enforcement and EMS personnel should develop a formal safety procedure for emergency rescue in an unprotected trench. Entering an unprotected trench after a cave-in or collapse could place would-be rescuers in danger. Rescue is a delicate and slow operation requiring knowledge of the behavior of unstable soil, necessary to prevent further injury to the victim or the rescuers. The added weight and vibrations can also contribute to an increased susceptibility to further collapse. Many rescuers precipitate second and third stage trench cave-ins and have become victims themselves. In this incident EMS personnel entered the unprotected trench in an attempt to rescue the victim, exposing themselves to an excavation collapse hazard.

Emergency rescue workers, such as law enforcement officials and EMS personnel, should receive specialized training in how to rescue workers who may be trapped in utility trenches, and should not put themselves in danger by entering an unprotected trench. In this incident, a specialized rescue team was called in to respond to the emergency. The rescue workers had special equipment for trench rescues and building collapses and had undergone specialized training in the area of trench/building collapse emergencies. They immediately constructed a wooden safety box in the trench with a system of ropes and pulleys before entering the trench to free the victim. National Fire Protection Association (NFPA) 1670, Chapter 11 details the requirements for rescue operations after a trench cave-in occurs.

Recommendation #6: *Local governing bodies and codes enforcement officers should receive additional training to upgrade their knowledge and awareness of high hazard work, including excavation and trenching. This skills upgrade should be provided to both new and existing codes enforcement officers.*

Discussion: This recommendation may create a mechanism of observation and oversight by the codes enforcement officers who are likely to encounter small employers and independent contractors during their work. The officers could inform the employers and contractors of potential hazards, provide fact sheets that highlight the key requirements for the excavation and trenching standards, and check some of the basics of the trenching project such as depth of the trench, protection of the trench and identification of the competent person. In addition, they could advise employers and contractors to contact safety experts to learn about and implement trench safety. This may be an effective accident prevention strategy, reaching the thousands of untrained and unprepared small employers and independent contractors with awareness and guidance, the very workers who represent the major group of fatalities in New York State.

In this incident, the town water and sewer inspector observed workers in the unprotected trench serving as spotters, observed a worker hand digging within a few feet of a live buried electrical utility, and observed the victim spotting in the unprotected trench for the excavating subcontractor while attempting to locate the sewer main. If the above recommendation was in place, with a trained and knowledgeable officer, at a minimum the excavation work may have been halted and entry into an unprotected trench may have been prohibited.

Recommendation #7: *Local governing bodies and codes enforcement officers should consider requiring building permit applicants to certify that they will follow written excavation and trenching plans in accordance with applicable standards and regulations, for any projects involving excavation and trenching work, before the building permits can be approved.*

Discussion: Local governing bodies may consider revising building permits to require building permit applicants to certify that they will follow written plans for any projects involving excavation and trenching. Statements on the permit applications would be added to indicate that the employer/independent contractor agrees to accept and abide by all standards and regulations governing the excavation and trenching work, not just local governing body codes and ordinances. If construction companies and independent contractors were required to provide written documentation of how the high hazard work of excavation and trenching will be performed safely as part of the building permit application process, it may prompt the employers and contractors to plan ahead, formally assess the hazards, seek assistance in developing the required safety and injury prevention program, and implement the necessary injury prevention measures. No work should be initiated unless these requirements are met after review and approval. These changes may help to prevent trench related fatalities in NYS.

Recommendation #8: *Employers and independent contractors should require that all employees are protected from exposure to electrical hazards in a trench.*

Discussion: Utilities to the single family residence were located underground in the trench near the edge of the road. Workers were observed using power and hand tools within inches of live 12,000 volt lines. This did not contribute to the fatality, but did present another potential hazard to workers in the excavation and trenching project and to the rescue workers. Performing cutting work next to hot utility lines could have resulted in additional serious injuries and death from electrocution. The company performed the utility mark-out as required by local codes but did not contact the utility company to turn off the power as required, when they realized the need to hand cut large rocks and boulders in the trench. The power was not shut off to these lines until after the incident, when workers returned to complete the work.

Key words: Trench, collapse, cave-in, trenching, excavation, trench protection systems, entrapment, spoils piles

REFERENCES:

1. Associated General Contractors of America Safety Training for the Focus Four. *Hazards in Construction.* Retrieved February 8, 2011 from http://www.agc.org/cs/career_development/safety_training/focus_four_locations

2. CDC/NIOSH. *NIOSH Safety and Health Topic: Trenching and Excavation.* Retrieved on February 8, 2011 from http://www.cdc.gov/niosh/topics/trenching/

3. CDC/NIOSH. MMWR. 2004. *Occupational Fatalities During Trenching and Excavation Work - United States, 1992-2001*. Morbidity and Mortality Weekly Report, 53(15):311-314. Retrieved February 8, 2011 from www.cdc.gov/mmwr/preview/mmwrhtml/mm5315a2.htm

4. CDC/NIOSH. Alert: July 1985. *Preventing Deaths and Injuries from Excavation Cave-ins*. retrieved February 8, 2011 from http://www.cdc.gov/niosh/docs/85-110

5. CDC/NIOSH. Fatality Assessment and Control Evaluation (FACE) investigation reports. Retrieved February 8, 2011 from www.cdc.gov/niosh/face

6. Center to Protect Workers' Rights (CPWR). Plog, Barbara et al. March, 2006. *Barriers to Trench Safety: Strategies to Prevent Trenching-Related Injuries and Deaths*. Retrieved February 8, 2011 from www.elcosh.org.

7. Commonwealth of Massachusetts. Executive Office of Labor and Workforce Development. *Trenching Hazard Alert for Public Works Employers and Employees in Massachusetts*. Bulletin 407, 11/2007, p1-4.

8. Deatherage, J.H., et al. 2004 *Neglecting Safety Precautions may lead to trenching fatalities*. American Journal of Industrial Medicine, 45(6):522-7.

9. EC&M online. June, 2009. *Danger Uncovered*. Beck, Ireland. Retrieved February 8, 2011 from http://ecmweb.com/construction/electrical-trench-safety-20090601/

10. Encyclopedia of Occupational Health and Safety. 4th Edition. *Chapter 93: Construction Trenching* by Jack Mickle. *Types of Projects and Their Associated Hazards* by Jeffrey Hinkman. Retrieved February 8, 2011 from
     http://www.elcosh.org/en/document/296/d000279/encyclopedia-of-occupational-safety-%2526-health-%253A-chapter-93-construction.html

11. Executive Safety Update. The Monthly News Bulletin of the Construction Safety Center, Vol. 17, Issue 3, September, 2009

12. Hinze, J.W. and K. Bren. 1997. *The causes of trenching-related fatalities*. Construction Congress V: Managing Engineered Construction in Expanding Global Markets. Proceedings of the Congress, sponsored by the American Society of Civil Engineers (ASCE), 131(4): 494-500.

13. Irizarry, J. et al: 2002 *Analysis of Safety Issues in Trenching Operation*. 10th Annual Symposium on Construction Innovation and Global Competitiveness, September 9-13, 2002. Retrieved February 8, 2011 from Construction Safety Alliance site:
     http://engineering.purdue.edu/CSA/publications/trenching03

14. Job Health and Safety Quarterly. Fall, 2009. *Trenching is a Dangerous and Dirty Business*. Retrieved February 8, 2011 from http://www.elcosh.org/en/document/161/d000168/trenching-is-a-dangerous-and-dirty-business.html

15. Miami-Dade County. *Trench Safety Act Compliance Statement, FM5238 Rev. (12-00)*. Retrieved February 8, 2011 from http://facilities.dadeschools.net/form_pdfs/5238.pdf

Slide 8

16. New York City Department of Buildings. *Excavation and Trench Safety Guidelines* by Dan Eschenasy. www.NYC.gov/buildings. Retrieved February 8, 2011 from http://www.elcosh.org/en/document/161/d000168/trenching-is-a-dangerous-and-dirty-business.html

17. OSHA. *Working Safely in Trenches Safety Tips.* Retrieved February 8, 2011 from http://www.osha.gov/Publictions/trench/trench_safety_tips_card.html

18. OSHA. *29CFR1926.650 subpart p. Excavations: scope, application and definitions.* Retrieved February 8, 2011 from http://www.osha.gov/pls/oshaweb/owadisp.show_document?p_id=10774&p_table=STANDARDS

19. OSHA. *29CFR1926.651 subpart p. Excavations: specific excavation requirements.* Retrieved February 8, 2011 from http://www.osha.gov/pls/oshaweb/owadisp.show_document?p_table=STANDARDS&p_id=10775

20. OSHA. *29CFR1926.652 subpart p. Excavations: requirements for protective systems.* Retrieved February 8, 2011 from http://www.osha.gov/pls/oshaweb/owadisp.show_document?p_table=STANDARDS&p_id=10776

21. OSHA. *OSHA Technical Manual SECTION V: CHAPTER 2 EXCAVATIONS: HAZARD RECOGNITON IN TRENCHING AND SHORING.* Retrieved February 8, 2011 from http://www.osha.gov/dts/osta/otm/otm_v/otm_v_2.html

22. OSHA. *OSHA's Construction e-tool.* Retrieved February 8, 2011 from http://www.osha.gov/SLTC/etools/construction/trenching/mainpage.html

The New York State Fatality Assessment and Control Evaluation (NY FACE) program is one of many workplace health and safety programs administered by the New York State Department of Health (NYSDOH). It is a research program designed to identify and study fatal occupational injuries. Under a cooperative agreement with the National Institute for Occupational Safety and Health (NIOSH), the NY FACE program collects information on occupational fatalities in New York State (excluding New York City) and targets specific types of fatalities for evaluation. NY FACE investigators evaluate information from multiple sources and summarize findings in narrative reports that include recommendations for preventing similar events in the future. These recommendations are distributed to employers, workers, and other organizations interested in promoting workplace safety. The NY FACE does not determine fault or legal liability associated with a fatal incident. Names of employers, victims and/or witnesses are not included in written investigative reports or other databases to protect the confidentiality of those who voluntarily participate in the program.

Additional information regarding the NY FACE program can be obtained from:
New York State Department of Health FACE Program
Bureau of Occupational Health
Flanigan Square, Room 230
547 River Street
1-518-402-7900
www.nyhealth.gov/nysdoh/face/face.htm

ACCIDENT SUMMARY No. 59

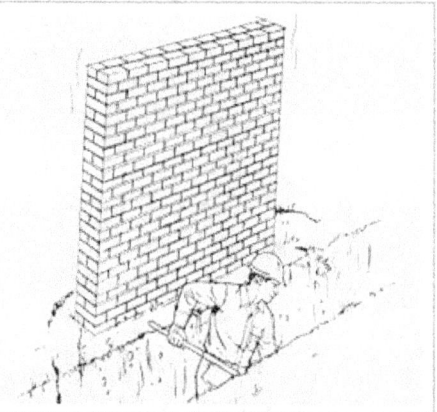

Accident Type:	Struck by Falling Wall
Weather Conditions:	Clear/Wet Soil
Type of Operation:	Trenching
Size of Work Crew:	2
Competent Safety Monitor on Site:	No
Safety and Health Program in Effect:	Inadeqaute
Was the Worksite Inspected Regularly:	No, short duration
Training and Education Provided:	Some
Employee Job Title:	Laborer
Age & Sex:	27-Male
Experience at this Type of Work:	1 Year
Time on Project:	1 Day

BRIEF DESCRIPTION OF ACCIDENT

An employee was in the process of locating an underground water line. A trench had been dug approximately 4 feet deep along side a brick wall 7 feet high and 5 feet long. The brick wall collapsed onto the victim who was standing in the trench. The injuries were fatal.

INSPECTION RESULTS

As a result of its investigation, OSHA issued citations for violation of the standard.

ACCIDENT PREVENTION RECOMMENDATIONS

The contractor should not permit employees to excavate below the level of the base of foundation footings when walls are unpinned [29 CFR 1926.651(i)(1)]

SOURCES OF HELP

- **OSHA 2202 Construction Industry Digest** ‾ includes all OSHA construction standards and those general industry standards that apply to construction. Order No. 029-016-00151-4, ($2.25). Available from the Superintendent of Documents, Government Printing Office, Washington DC 20402-9325, phone (202) 512-1800. Make checks payable to Superintendent of Documents. For phone orders, Visa® or MasterCard®.
- **OSHA 2254 Training Requirements in OSHA Standards and Training Guidelines** ‾ includes all OSHA construction standards and those general industry standards that apply to construction. Order No. 029-016-00160-3, ($6.00). Available from the Superintendent of Documents, Government Printing Office, Washington DC 20402-9325, phone (202) 512-1800. Make checks payable to Superintendent of Documents. For phone orders, Visa® or MasterCard®.
- **OSHA Safety and Health Guidelines for Construction** (Available from the National Information Service, 5285 Port Royal Road, Springfield, VA 22161; (703) 605-6000 or (800) 553-6847; Order No. PB-239-312/AS, $27). Guidelines to helpconstruction employers establish a training program in the safe use of equipment, tools, and machinery on the job.

- For information on OSHA-funded free consultation services call the nearest OSHA area office listed in telephone directories under U.S. Labor Department or under the state government section where states administer their own OSHA programs.
- Courses in construction safety are offered by the OSHA Training Institute, 1555 Times Drive, Des Plaines, IL 60018, 708/297-4810.
- OSHA Safety and Health Training Guidelines for Construction (Available from the National Technical Information Service, 5285 Port Royal Road, Springfield, VA 22161; 703/487-4650; Order No. PB-239-312/AS): guidelines to help construction employers establish a training program in the safe use of equipment, tools, and machinery on the Job.

NOTE: The case here described was selected as being representative of fatalities caused by improper work practices. No special emphasis or priority is implied nor is the case necessarily a recent occurrence. The legal aspects of the incident have been resolved, and the case is now closed.

Scaffolding Accidents

Design as a Risk Factor: Australian Study, 2000–2002

- Main finding: design contributes significantly to work-related serious injury.

- 37% of workplace fatalities are due to design-related issues.

- In another 14% of fatalities, design-related issues may have played a role.

[Driscoll et al. 2008]

Photo courtesy of Thinkstock

CDC NIOSH

Structural Steel

NOTES

Several studies around the world have demonstrated that design can directly affect the safety of a construction site or process. The Australian government investigated the design-related root causes of their work-related fatalities. Seventy-seven (37%) of the 210 identified workplace fatalities definitely or probably had design-related issues involved. In another 29 fatalities (14%), the circumstances suggested that design issues were involved. The most common scenarios involved problems with rollover protective structures and/or associated seat belts; inadequate guarding; lack of residual current devices; inadequate fall protection; failed hydraulic lifting systems in vehicles and mobile equipment; and inadequate protection mechanisms on mobile plants and vehicles.

These fatal incidents might have been prevented if the hazards that caused them had been considered during the design phase.

SOURCES

Driscoll TR, Harrison JE, Bradley C, Newson RS [2008]. The role of design issues in work-related fatal injury in Australia. J Safety Res 39(2):209–14 [Epub 2008:Mar 13; PubMed index for MEDLINE: 18454972].

NIOSH Fatality Assessment and Control Evaluation (FACE) Program [1983]. Fatal incident summary report: scaffold collapse involving a painter. FACE 8306 [www.cdc.gov/niosh/face/In-house/full8306.html].

Photo courtesy of Thinkstock

Fatal Incident Summary Report: Scaffold Collapse Involving a Painter

INTRODUCTION

The National Institute for Occupational Safety and Health (NIOSH), Division of Safety Research (DSR), is currently conducting the Fatal Accident Circumstances and Epidemiology (FACE) Study. By scientifically collecting data from a sample of similar fatal accidents, this study will identify and rank factors which increase the risk of fatal injury for selected employees.

On May 25, 1983, a painter suffered fatal injuries when the suspended scaffolding from which he was working collapsed. The County Coroner requested NIOSH technical assistance to develop information on factors involved with the incident data.

CONTACTS/ACTIVITIES

After receiving notification, three Division of Safety Research personnel, a safety specialist, a safety engineer, and an epidemiologist, visited at the site to interview the employer and witnesses and to obtain comparison data from suitable co-workers. The research team, the police department, and the employer examined the impounded scaffold at an independent testing laboratory.

A debriefing session was held with the employer, other employees, and the contractor. During this introductory meeting, background information was obtained about the contractor and the employer, including an overview of their safety and health program. Interviews were conducted with witnesses and co-workers. Examining the scaffold assisted the researchers in developing hypotheses about the sequence of events leading to the incident.

SYNOPSIS OF EVENTS

The two workers had placed the scaffold supporting wire rope on the 7th floor permanently installed eye hooks. They then reeved the wire rope to the scaffold stirrups which are located at each end of the scaffold staging. After reeving was complete, the workers raised the scaffolding to the 7th floor windows. This action was accomplished by turning the drive motor directional switch to the "up" position and holding the motor switch in the "on" position.

The victim had to apply caulking around the windows. After caulking half way across the floor, he had to change positions, including independent life lines with a co-worker, who survived the incident. After caulking the remaining windows, the workers switched positions again in order to begin their descent.

The co-worker stated that he turned away from the victim and faced his stirrup in preparation of descent. As he did this, he felt some movement in the scaffold. He turned and looked at the victim, who motioned by hand signal to turn the directional switch to the "down" position. The co-worker signaled "okay" and turned to face his stirrup. As he was in the process of preparing his stirrup for downward movement plus getting his lanyard grab device ready to move down, he felt several sudden jerks and was suddenly dangling from his life line. After regaining his composure, the co-worker looked for the victim in the area of his life line. The co-worker then noticed the victim lying in the street across from the building.

GENERAL CONCLUSIONS AND RECOMMENDATIONS

There is some evidence which indicates the deceased was not familiar with the operation of this type of scaffold. For this type of scaffold, the operator must operate the drill and a brake lever at the same time with one hand, while releasing his lanyard on the safety line with the other hand.

Additionally, the victim's lanyard failed to prevent the fatal fall for one of two reasons. Either the lanyard was deteriorated to the extent that the impact load was in excess of the lanyard strength or the lanyard became entangled in the scaffold components.

It is suspected that the wire rope broke because the hoist's secondary safety mechanism did not function quickly enough. The wire rope broke at a level 20+ feet below where the scaffold was originally positioned. When the mechanism finally activated, the force of the falling scaffold caused the emergency braking cam to squeeze the rope to such an extent that it actually cut 5 of the 6 strands. The remaining strand was not of sufficient strength to hold the falling scaffold and it also broke.

It is recommended that workers who use scaffolds should be trained in the proper use, maintenance, and limitations of scaffolding, life lines and lanyards. Also management should be aware of their responsibilities when their workers are using scaffolds. Safety requirements for scaffolding are outlined in the OSHAct regulations 1910.28, 1910.29 and 1926.451.

Accidents Linked to Design

- 22% of 226 injuries that occurred from 2000 to 2002 in Oregon, Washington, and California were linked partly to design [Behm 2005]

- 42% of 224 fatalities in U.S. between 1990 and 2003 were linked to design [Behm 2005]

- In Europe, a 1991 study concluded that 60% of fatal accidents resulted in part from decisions made before site work began [European Foundation for the Improvement of Living and Working Conditions, 1991]

- 63% of all fatalities and injuries could be attributed to design decisions or lack of planning [NOHSC 2001]

NOTES

Research conducted in the United States, Europe, and other regions has shown that design does affect the inherent risk in constructing a facility. Research linked design to 22% of injuries that occurred in western states and 42% of fatalities across the country. European researchers found that nearly two-thirds of fatalities and injuries were linked to design. Facility designers are encouraged to consult with occupational safety and health professionals early in the design process to identify and design out hazards and to reduce risk of injury, illness, and death.

SOURCES

Behm M [2005]. Linking construction fatalities to the design for construction safety concept. Safety Sci 43:589–611.

NOHSC [2001]. CHAIR safety in design tool. New South Wales, Australia: National Occupational Health & Safety Commission.

European Foundation for the Improvement of Living and Working Conditions [1991]. From drawing board to building site (EF/88/17/FR). Dublin: European Foundation for the Improvement of Living and Working Conditions.

Falls

 **Falls**

- Number one cause of construction fatalities
  - in 2010, 35% of 751 deaths
    www.bls.gov/news.release/cfoi.t02.htm

- Common situations include making connections, walking on beams or near openings such as floors or windows

- Fall protection is required at height of 6 feet above a surface [29 CFR 1926.760].

- Common causes: slippery surfaces, unexpected vibrations, misalignment, and unexpected loads

NOTES

Falls are the number one cause of deaths in the construction industry. In 2004, 445 (36%) of 1,234 deaths were due to falls [BLS 2006]. By contrast, of 751 deaths in the construction sector in 2010, 35% were attributed to falls [BLS 2011a]. The decline in number of fatalities in the construction sector in 2010, compared to 2004, was attributed more to the economic downturn than to any other factor [BLS 2011b].

Falls from any height can be fatal. In construction, workers are often high off the ground. For structural reasons, the taller cross-sections of W shapes are usually chosen for beams. The flanges on W shapes may be less than six inches wide. Workers walk on beams, sometimes without fall protection. Fall protection is highly recommended and often required in most scenarios involving heights. OSHA requires fall protection at a height of 15 feet above a surface during steel erection. For other construction phases, it is 6 feet [29 CFR 1926.760].

SOURCES

BLS [2006]. Injuries, illnesses, and fatalities in construction, 2004. By Meyer SW, Pegula SM. Washington, DC: U.S. Department of Labor, Bureau of Labor Statistics, Office of Safety, Health, and Working Conditions [www.bls.gov/opub/cwc/sh20060519ar01p1.htm].

BLS [2011a]. Census of Fatal Occupational Injuries. Washington, DC: U.S. Department of Labor, Bureau of Labor Statistics. [www.bls.gov/news.release/cfoi.t02.htm]

BLS [2011b]. Injuries, Illnesses, and Fatalities (IIF). Washington, DC: U.S. Department of Labor, Bureau of Labor Statistics. [www.bls.gov/iif/home.htm]

OSHA [2001]. Standard number 1926.760: fall protection. Washington, DC: U.S. Department of Labor, Occupational Safety and Health Administration.

 **Death from Injury**

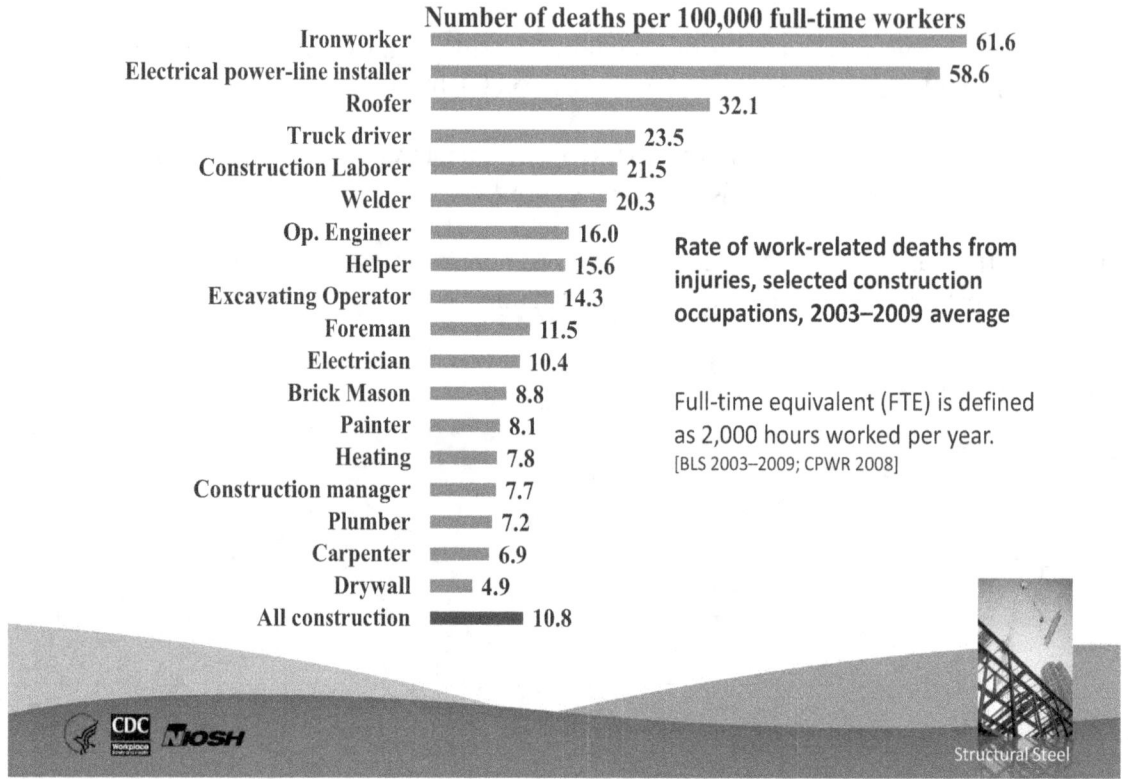

Number of deaths per 100,000 full-time workers

Ironworker	61.6
Electrical power-line installer	58.6
Roofer	32.1
Truck driver	23.5
Construction Laborer	21.5
Welder	20.3
Op. Engineer	16.0
Helper	15.6
Excavating Operator	14.3
Foreman	11.5
Electrician	10.4
Brick Mason	8.8
Painter	8.1
Heating	7.8
Construction manager	7.7
Plumber	7.2
Carpenter	6.9
Drywall	4.9
All construction	10.8

Rate of work-related deaths from injuries, selected construction occupations, 2003–2009 average

Full-time equivalent (FTE) is defined as 2,000 hours worked per year.
[BLS 2003–2009; CPWR 2008]

Structural Steel

NOTES

The Center for Construction Research and Training compiles a "Construction Chart Book" using Bureau of Labor Statistics data [CPWR 2008]. It includes two illuminating charts useful for considering safety issues. This chart is compiled from 2003–2009 data on workplace fatalities. Ironworkers experience the highest work-related death rate, with 61.6 fatalities per 100,000 FTE.

SOURCES

BLS [2003–2009]. Census of Fatal Occupational Injuries. Washington, DC: U.S. Department of Labor, Bureau of Labor Statistics [www.bls.gov/iif/oshcfoi1.htm].

CPWR [2008]. The construction chart book. 4th ed. Silver Spring, MD: Center for Construction Research and Training.

Fatality Assessment and Control Evaluation

NIOSH FACE Program www.cdc.gov/niosh/face

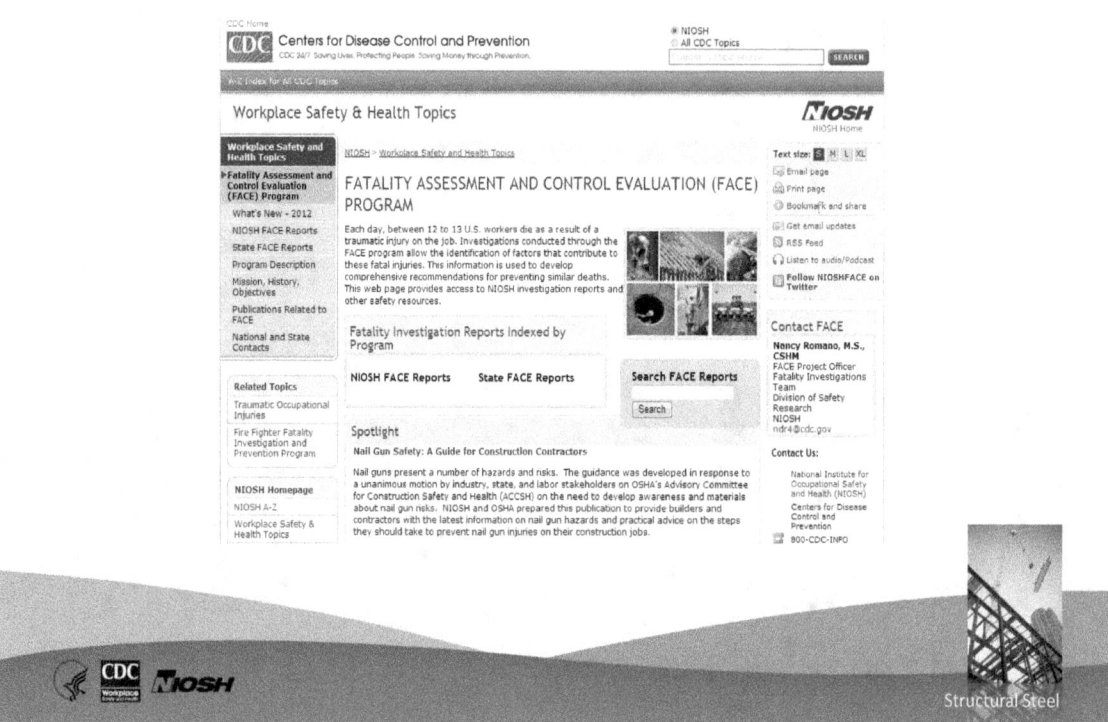

NOTES

The NIOSH Fatality Assessment and Control Evaluation Program examines worker fatalities by type of injury. By studying these reports, an enterprising designer can identify recurrent problems to "design out."

SOURCE

NIOSH Fatality Assessment and Control Evaluation Program [www.cdc.gov/niosh/face/]

What is Prevention through Design?

Eliminating or reducing work-related hazards and illnesses and minimizing risks associated with

- Construction

- Manufacturing

- Maintenance

- Use, reuse, and disposal of facilities, materials, and equipment

NOTES

PtD is a risk management technique that is being applied successfully in many industries, including manufacturing, healthcare, telecommunications, and construction. PtD is the optimal method of preventing occupational illnesses, injuries, and fatalities by designing out the hazards and risks. This approach involves the design of tools, equipment, systems, work processes, and facilities in order to reduce, or eliminate, hazards associated with work. The concept is simply that the safety and health of workers throughout the life cycle are considered while the product and/or process is being designed. The life cycle starts with concept development, and includes design, construction or manufacturing, operations, maintenance, and eventual disposal of whatever is being designed, which could be a facility, a material, or a piece of equipment.

PtD processes have been required in other countries for several years now, but in the United States PtD is being adopted on a voluntary basis. The National Institute for Occupational Safety and Health (NIOSH) is spearheading a national initiative in PtD and partnering with many professional organizations to apply the concept to their industry and professions. The Occupational Safety and Health Administration (OSHA) is very interested in PtD but is not currently considering making it mandatory.

PtD design professionals (that is, architects and/or engineers) working with the project owner (that is, the client) make deliberate design decisions that eliminate or reduce the risk of injuries or illness throughout the life of a project, beginning at the earliest stages of a project's life cycle. PtD is thus the deliberate consideration of construction and maintenance worker safety and health in the design phase of a construction project. PtD processes in construction have been required in the United Kingdom for over a decade and are being implemented in other countries such as Australia and Singapore.

PtD applies to the design of a facility, that is, to the aspects of the completed building that make a project inherently safer. PtD does not focus on how to make different methods of construction safer. For example, it does not focus on how to use fall protection systems, but it does include consideration of design decisions that influence how often fall protection will be needed. Similarly, PtD does not address how to erect safe scaffolding, but it does relate to design decisions that influence the location and type of scaffolding needed to accomplish the work. PtD concepts may also be used to design temporary structures. Some design decisions improve workplace safety. For example, when the height of parapet walls is designed to be 42", the parapet acts as a guardrail and enhances safety. When designed into the permanent structure of the building and sequenced early in construction, the parapet at this height acts to enhance safety during initial construction activities and during subsequent maintenance and construction activities, such as roof repair. In the United States, the employer is solely responsible for site safety.

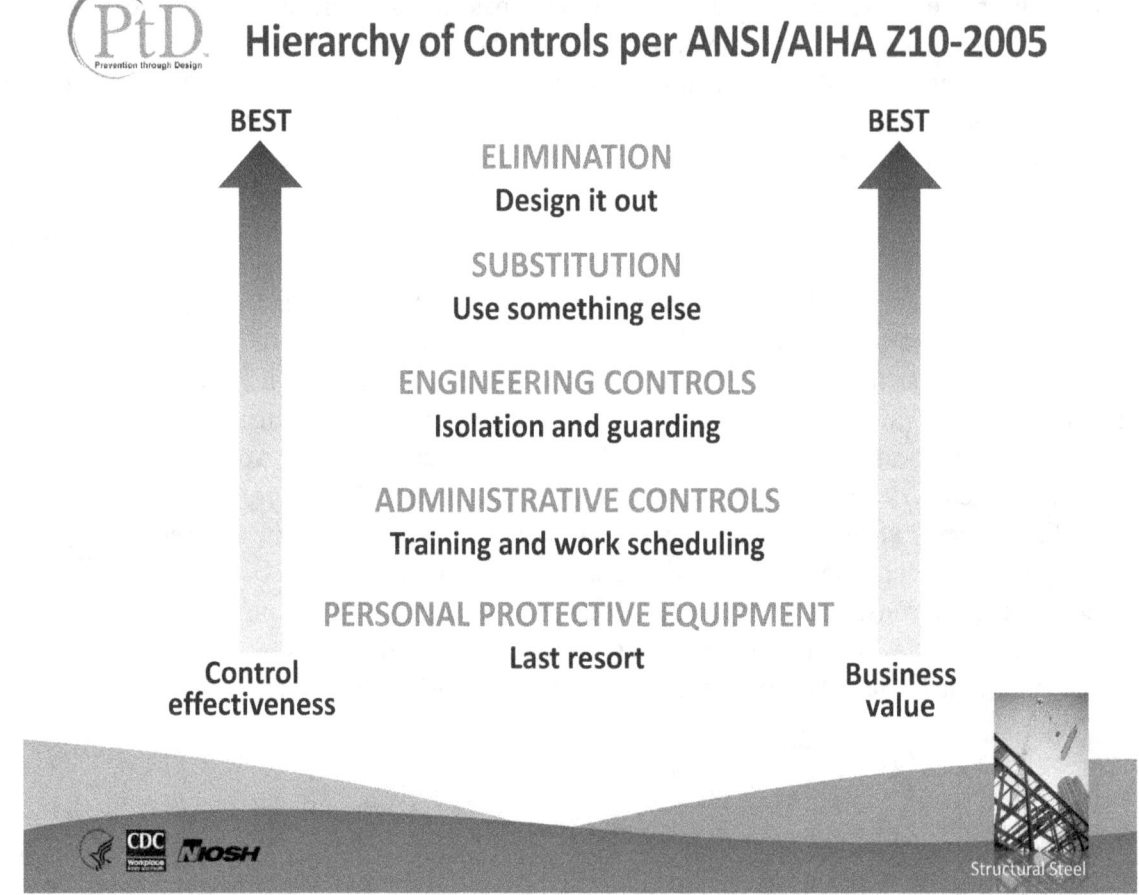

Hierarchy of Controls per ANSI/AIHA Z10-2005

BEST

BEST

ELIMINATION
Design it out

SUBSTITUTION
Use something else

ENGINEERING CONTROLS
Isolation and guarding

ADMINISTRATIVE CONTROLS
Training and work scheduling

PERSONAL PROTECTIVE EQUIPMENT
Last resort

Control
effectiveness

Business
value

Structural Steel

NOTES

This slide shows the well-accepted Hierarchy of Controls. PtD anticipates and removes potential hazardous elements at the design phase of a project through elimination or substitution. Residual risks may be minimized through the use of engineering and administrative controls.

The top of the hierarchy is better in terms of improved occupational safety and health (OSH) and cost savings. Below is a description of the different levels, from most to least effective.

Elimination: "Design out" hazards and hazardous exposures.

Substitution: Substitute less-hazardous materials, processes, operations, or equipment. A larger crane may be specified when the load or the reach approaches the crane design limit. Nontoxic chemicals are preferred. The Green Chemistry movement replaces toxic compounds with less hazardous chemicals.

Engineering controls: Isolate process or equipment or contain the hazard. Remove hazard from work zone, e.g., with exhaust ventilation. Require two hands to operate machinery. Use warning devices to warn worker about entry into hazard zone. Signs, labels, alarms, and flashing lights give warnings. Safety switches, hand guards, and other engineering controls prevent certain kinds of injuries.

Administrative controls: Job rotation, work scheduling, training, well-designed work methods, and organization are examples. Administrative controls include training modules and company procedures. A well-organized worksite is safer than a messy one. Reducing the clutter on a construction site improves worker safety by reducing the exposure to hazards. The foreman controls site layout and housekeeping policies.

Personal Protective Equipment (PPE): Includes but is not limited to safety glasses for eye protection; ear plugs for hearing protection; clothing such as safety shoes, gloves, and overalls; face shields for welders; fall harnesses; and respirators to prevent inhalation of hazardous substances.

SOURCE

ANSI/AIHA [2005]. American national standard for occupational health and safety management systems. New York: American National Standards Institute, Inc. ANSI/AIHA Z10-2005.

Personal Protective Equipment (PPE)

- Last line of defense against injury
- Examples:
  - Hard hats
  - Steel-toed boots
  - Safety glasses
  - Gloves
  - Harnesses

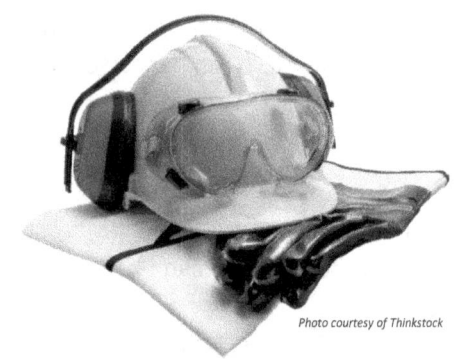

Photo courtesy of Thinkstock

OSHA www.osha.gov/Publications/osha3151.html

Structural Steel

NOTES

PPE includes items worn as a last line of defense against injury. OSHA-required PPE can include hardhats, steel-toed boots, safety glasses or safety goggles, gloves, earmuffs, full body suits, respiratory aids, face shields, and fall harnesses.

SOURCES

NOHSC [2001]. CHAIR safety in design tool. New South Wales, Australia: National Occupational Health & Safety Commission.

OSHA PPE publications

www.osha.gov/Publications/osha3151.html

www.osha.gov/OshDoc/data_General_Facts/ppe-factsheet.pdf

www.osha.gov/OshDoc/data_Hurricane_Facts/construction_ppe.pdf

Photo courtesy of Thinkstock

PtD Process

[Hecker et al. 2005]

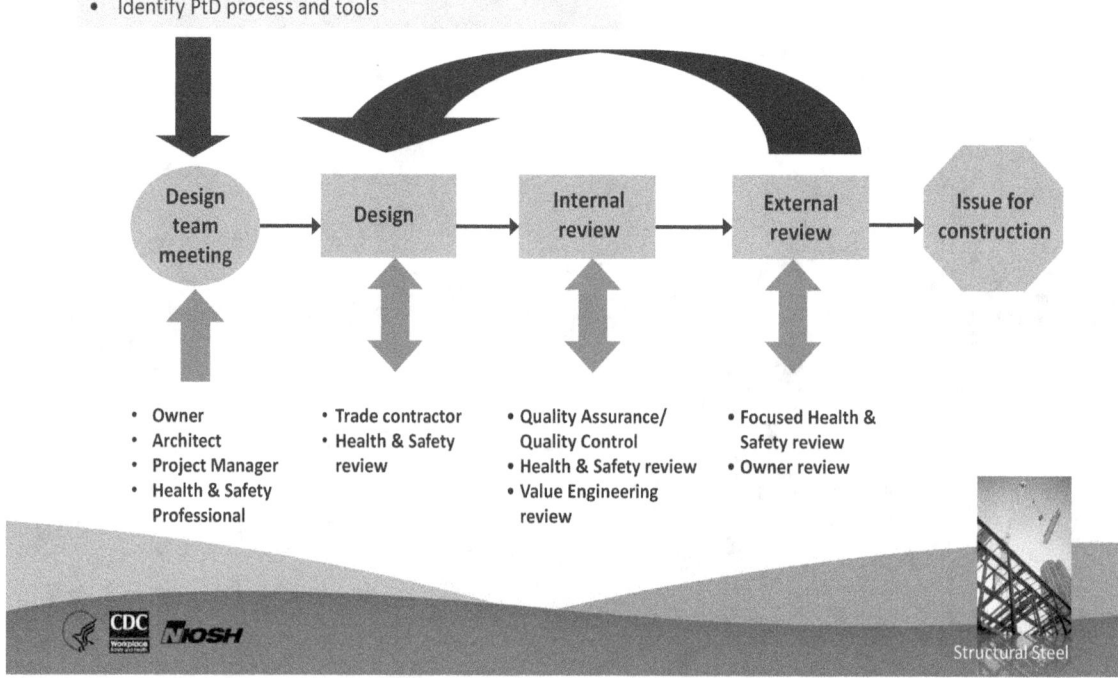

NOTES

This graphic depicts the typical PtD process. The key component of this process is the incorporation of safety knowledge into design decisions. For example, site safety should be considered throughout the design process. A progress review specifically focused on site safety may be effective. Site safety knowledge can be provided by trade contractors, an on-site employee, or a hired consultant. The graphic emphasizes the importance of communication between designers and constructors. Such communication during design may reveal steps to reduce construction duration.

Many Project Managers schedule a Value Engineering review prior to issuing drawings for bid. The purpose is to reduce overall project costs. Unfortunately, during the review, redundant systems that are necessary to protect worker health may be eliminated. It is therefore considered a best practice to conduct a focused Health & Safety (H&S) review before drawings are issued.

SOURCE

Hecker S, Gambatese J, Weinstein M [2005]. Designing for worker safety: moving the construction safety process upstream. Prof Saf *50*(9):32–44.

 **Integrating Occupational Safety and Health with the Design Process**

Stage	Activities
Conceptual design	Establish occupational safety and health goals, identify occupational hazards
Preliminary design	Eliminate hazards, if possible; substitute less hazardous agents/processes; establish risk minimization targets for remaining hazards; assess risk; and develop risk control alternatives. Write project specifications.
Detailed design	Select controls; conduct process hazard reviews
Procurement	Develop equipment specifications and include in procurements; develop "checks and tests" for factory acceptance testing and commissioning
Construction	Ensure construction site safety and contractor safety
Commissioning	Conduct "checks and tests," including factory acceptance; pre–start up safety reviews; development of standard operating procedures (SOPs); risk/exposure assessment; and management of residual risks
Start up and occupancy	Education; manage changes; modify SOPs

NOTES

The integration of OSH goals within the design processes is an essential concept because it elevates the importance of safety and health as a value proposition in the overall design, construction, and operation of projects.

Identify hazards during conceptual design. Follow the Hierarchy of Controls to eliminate or reduce risks.

For example, how much space is needed to access, maintain, and replace HVAC units?

Use project specifications to require the inclusion of fall protection systems such as permanent anchor points for lifelines. Reduce fall hazards by specifying a ladder-free construction site.

Obtain a site plan that shows the location of existing underground and overhead utilities and develop traffic control plans to avoid those hazards.

Compare the list of desirable safety features against the detailed design.

Obtain feedback from safety and health professionals, contractors, and trade representatives. Modify the design to improve safety.

Call out required hazard controls on the drawing and in the contract specifications when possible. During procurement, compare materials and equipment received against the contract specifications. Develop a checklist for commissioning.

During construction, how do contractors communicate with the project manager and each other? Who has the authority to correct a hazardous condition on the worksite?

What procedures are followed before and after permanent equipment reaches the site? Follow the commissioning checklist!

Does the building have unusual features? Educate the owners and tenants.

Are special operating procedures required?

At each stage of the design process, think of ways to reduce the workplace risks.

 **Safety Payoff During Design**

[Adapted from Szymberski 1997]

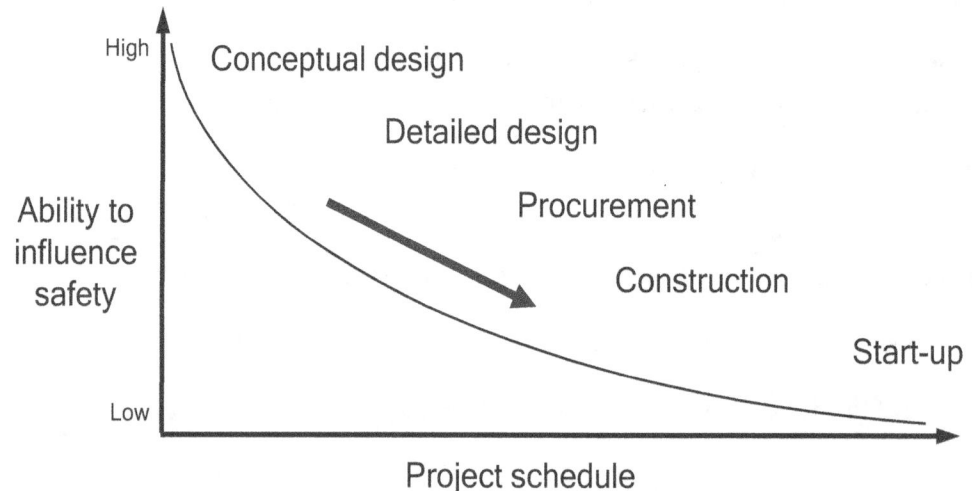

NOTES

Most owners and design professionals know intuitively that the earlier in the design process that cost is considered, the easier it is to achieve cost-effective goals. The same is true for construction duration and quality. A worker's ability to influence project criteria decreases as the design and construction progress. The same principle is true for construction safety. The earlier in the project life cycle that safety is considered, the easier it is to reduce hazards. This concept is in contrast to the prevailing methods of planning for construction site safety, which do not begin until a short time before the construction phase, when the ability to influence safety is limited.

SOURCE

Szymberski R [1997]. Construction project planning. TAPPI J *80*(11):69–74.

PtD Process Tasks

[Adapted from Toole 2005; Hinze and Wiegand 1992]

- Perform a hazard analysis

- Incorporate safety into the design documents

- Make a CAD model for member labeling and erection sequencing

Photo courtesy of Thinkstock

Structural Steel

NOTES

This slide provides more details about the PtD process. Before, during, or after the conceptual design of a building, a hazard analysis can be performed. The designer meets with field professionals to review constructability, looking through the entire design for any hazards and addressing those hazards. The field professional can teach an inexperienced designer how to minimize risks in the field.

The safety input received during conceptual design can be reflected in detailed design drawings and specifications. Another constructability review should occur as the detailed design nears completion.

Sometimes the drawings that result from a PtD process look the same as typical construction drawings, but they are inherently safer for construction. Other times, drawings include special details and labels to make it easier for workers to erect the design safely.

Construction documents can be supplemented with graphic models and tables that contribute to safe erection. For example, a CAD file can be used to label steel members for safe erection sequencing. New software such as building information modeling (BIM) is able to show the final layouts of buildings and can detect any spatial problems before construction starts. Clearly labeled shop drawings eliminate confusion during installation. The BIM program can recommend efficient, safer erection sequencing.

SOURCES

Hinze J, Wiegand F [1992]. Role of designers in construction worker safety. Journal of Construction Engineering and Management *118*(4):677–684.

Toole TM [2005]. Increasing engineers' role in construction safety: opportunities and barriers. Journal of Professional Issues in Engineering Education and Practice *131*(3):199–207.

Photo courtesy of Thinkstock

Designer Tools

- Checklists for construction safety [Main and Ward 1992]

- Design for construction safety toolbox [Gambatese et al. 1997]

- Construction safety tools from Australia
  - Construction Hazard Assessment Implication Review, known as CHAIR [NOHSC 2001]

Structural Steel

NOTES

Most designers are not trained in PtD or construction site safety. It is therefore critical that they be given tools to facilitate the process. A PtD checklist alerts designers to common design elements that can lead to unnecessary hazards and identifies design options that are inherently safer. An example checklist is provided on the next slide.

The Design for Construction Safety Toolbox was developed by a Construction Industry Institute–sponsored research team that included leading PtD academics. This Toolbox was recently updated by Professor Jimmie Hinze at the University of Florida. The United Kingdom and Australia make available on the Web valuable PtD tools that reflect their experiences with PtD legislation and voluntary initiatives. For example, CHAIR (Construction Hazard Assessment Implication Review) is an Australian tool and methodology that systematically combines brainstorming and decisions to gradually rid the design of unnecessary hazards.

SOURCES

NOHSC [2001]. CHAIR safety in design tool. New South Wales, Australia: National Occupational Health & Safety Commission.

Gambatese JA, Hinze J, Haas CT [1997]. Tool to design for construction worker safety. J Arch Eng 3(1):2–41.

Main BW, Ward AC [1992]. What do engineers really know and do about safety? Implications for education, training, and practice. Mechanical Engineering 114(8):44–51.

Example Checklist

Item	Description
1.0	**Structural Framing**
1.1	Space slab and mat foundation top reinforcing steel at no more than 6 inches on center each way to provide a safe walking surface.
1.2	Design floor perimeter beams and beams above floor openings to support lanyards.
1.3	Design steel columns with holes at 21 and 42 inches above the floor level to support guardrail cables.
2.0	**Accessibility**
2.1	Provide adequate access to all valves and controls.
2.2	Orient equipment and controls so that they do not obstruct walkways and work areas.
2.3	Locate shutoff valves and switches in sight of the equipment which they control.
2.4	Provide adequate head room for access to equipment, electrical panels, and storage areas.
2.5	Design welded connections such that the weld locations can be safely accessed.

Checklist courtesy of John Gambatese

NOTES

Like many PtD checklists, this example includes hazards associated with both construction and maintenance.

SOURCE

Checklist courtesy of John Gambatese

OSHA Steel Erection eTool

OSHA www.osha.gov/SLTC/etools/steelerection/index.html

NOTES

To help steel design and construction employers determine what their employee safety responsibilities are, OSHA maintains Web pages that contain interpretations and clarifications of the federal standards and provide access to eTools that may be adopted voluntarily. This slide shows the steel erection eTool that delineates the requirements of Subpart R.

SOURCE

OSHA [www.osha.gov/SLTC/etools/steelerection/structural.html]

Why Prevention through Design?

- Ethical reasons

- Construction dangers

- Design-related safety issues

- Financial and non-financial benefits

- Practical benefits

Photo courtesy of Thinkstock

Structural Steel

NOTES

Engineers have strong ethical reasons to apply the PtD concept to their designs. There are practical benefits, too. Lost-time accidents delay the job, destroy crew morale, and cost money. The next few slides will show there are many reasons why owners and design professionals should be motivated to incorporate PtD in a project.

SOURCE

Photo courtesy of Thinkstock

Ethical Reasons for PtD

- National Society of Professional Engineers' Code of Ethics:

  "Engineers shall hold paramount the safety, health, and welfare of the public..."

- American Society of Civil Engineers' Code of Ethics:

  "Engineers shall recognize that the lives, safety, health and welfare of the general public are dependent upon engineering decisions..."

  NSPE www.nspe.org/ethics

  ASCE www.asce.org/content.aspx?id=7231

NOTES

Many safety professionals and design professionals believe that PtD is clearly an ethical duty. Nearly all national engineering societies include in their code of ethics a statement similar to the one shown here for the National Society of Professional Engineers: "Engineers shall hold paramount the safety, health, and welfare of the public."

The American Society of Civil Engineers goes one step further and explicitly states that engineering decisions directly affect safety. These organizations pledge to protect "the public." Why? The public lacks the knowledge of forces, stresses, and other risk-related issues that contribute to hazardous work-related conditions. Many construction and maintenance workers, especially apprentices, fail to perceive an unsafe condition. Even if construction workers recognize a hazard that could have been eliminated or reduced through an alternative design, there are significant barriers to redesign after construction is under way. Their safety and health deserve consideration.

SOURCES

American Society of Civil Engineers [ASCE][www.asce.org/Content.aspx?id=7231]

National Society of Professional Engineers [NSPE][www.nspe.org/ethics]

PtD Applies to Constructability

- How reasonable is the design?
  - Cost
  - Duration
  - Quality
  - Safety

Photo courtesy of the Cincinnati Museum Center www.cincymuseum.org

Structural Steel

NOTES

Most designers know that what may look great on paper might not be constructible. An important part of the design process is to evaluate the design's constructability, that is, to what extent the design can be constructed at a reasonable price, quickly, and with high quality. Safety is an important part of constructability. Accidents cost money, delay construction, and may result in bad publicity rather than acclaim for the owner.

Exciting buildings designed by creative architects require strong consideration of worker safety and health early in the design process. Owners realize these one-of-a-kind structures cost more to build and generally present unique challenges for the construction crew. Fewer construction firms have the expertise needed to build the structure, so fewer firms submit a bid, which reduces competition and therefore drives up price, resulting in higher bond and insurance costs. The timeline for procurement and construction is harder to estimate. The uniqueness of the design creates construction and maintenance challenges. Unusual materials, custom fabrications, non-standard specifications, and striking aesthetic features inherent in these designs require greater collaboration. The PtD process shown on the next slide helps the design team identify potential hazards in time to devise appropriate prevention strategies for construction crews and future maintenance workers. The project manager should include occupational safety and health professionals throughout the design process to design-in protections for workers.

SOURCE

Photo courtesy of the Cincinnati Museum Center

 **Business Value of PtD**

- Anticipate worker exposures—be proactive

- Align health and safety goals with business goals

- Modify designs to reduce/eliminate workplace hazards in

Facilities	Equipment
Tools	Processes
Products	Work flows

 Improve business profitability!

AIHA www.ihvalue.org

Structural Steel

NOTES

Companies that have implemented PtD programs experience lower than average injury and illness rates and lower workers' compensation expenses. However, the business value of PtD does not end there. In a study entitled Demonstrating the Business Value of Industrial Hygiene (known as The Value Study), findings showed that significant business cost savings accrue when hazards are eliminated or reduced.

SOURCE

American Institute of Industrial Hygienists [AIHA] [2008]. Strategy to demonstrate the value of industrial hygiene [www.aiha.org/votp_NEW/pdf/votp_exec_summary.pdf].

Benefits of PtD

- Reduced site hazards and thus fewer injuries
- Reduced workers' compensation insurance costs
- Increased productivity
- Fewer delays due to accidents
- Increased designer-constructor collaboration
- Reduced absenteeism
- Improved morale
- Reduced employee turnover

NOTES

PtD yields better value for owners and better health for the workers. When a project is designed with construction worker safety in mind, there are fewer hazards on site, with fewer injuries and fatalities. A reduction in injuries results in reduced workers' compensation insurance and less down-time, a direct savings for the employer. Experience shows PtD increases productivity and reduces labor costs. Safer designs lead to fewer project delays.

 **Industries Use PtD Successfully**

- Construction companies

- Computer and communications corporations

- Design-build contractors

- Electrical power providers

- Engineering consulting firms

- Oil and gas industries

- Water utilities

  And many others

NOTES

Major corporations in diverse industries and public utilities in several states have applied PtD through initiatives or established programs. At these companies, worker safety and health are an integral part of the corporate culture. International construction firms first encountered PtD on their European projects. They brought the concepts and related cost savings home to their American operations. Many firms provide PtD training for their design engineers in the areas of construction site safety, PtD checklists, and safety constructability reviews. These firms want to hire engineers who have a basic understanding of PtD.

Design, Detailing, and Fabrication Process

STRUCTURAL STEEL DESIGN

Design, Detailing, and Fabrication Process

Structural Steel

NOTES

This section summarizes the steps from design through fabrication.

 **Three Entities Associated with Design**

- Engineer
- Detailer
- Fabricator

Photo courtesy of Thinkstock

NOTES

Three entities participate in the design of structural steel: the engineer, the detailer, and the fabricator. The engineer creates the overall structural design, calculates loads according to building codes, determines the layout and sizes of structural steel members, and often designs the connections between members. The steel detailer takes contract documents and transfers relevant information to the shop drawings. In some cases, especially on the East Coast of the United States, the engineer will delegate the task of connection design to the steel detailer. This is not allowed on the West Coast, because the seismic activity in the area requires stricter design standards. The American Institute of Steel Construction (AISC) regulations dictate how certain aspects of connection design are managed before being delegated, such as load calculations. Sometimes the steel detailer's professional engineer will stamp the connection drawings. The steel fabricator completes the shop and erection drawings and uses them to fabricate steel members. The use of building infomation modeling (BIM) software by all parties facilitates more efficient collaboration and results in fewer errors.

SOURCE

Photos courtesy of Thinkstock

Design Phase

- Owner establishes architectural/engineering requirements for building

- Designer runs analysis on design according to building codes

- Building is designed for safety, serviceability, constructability, and economy

- Client receives final design specifications and drawings

- Designer stores the calculations

NOTES

The design process begins when an owner decides to build a facility and identifies an architect or engineer to design it. In most instances, an architect will be in charge of the main design, but an engineer will be in charge of the structural system. The owner will communicate necessary design criteria to the architect. The engineer will perform design and analysis based on building codes, such as the International Building Code (IBC). Four common design criteria are the following:

- Occupant safety—ensuring the structure will not fail and fall down under design loads
- Serviceability—ensuring that the structure performs well enough for use as it was intended and does not, for example, result in excessive vibration or beam deflection
- Economy—ensuring the building will not have excessive cost
- Constructability—ensuring that the structure can be built without excessive construction difficulties and includes construction safety.

The design will require multiple iterations. Once the designer feels the structure is as good as it can be, he or she gives the client the final design specifications and drawings. It is important to save neat, clear calculations as evidence of a good design.

 **Detailing**

Fabricator programs engineer's drawings with software to visualize connections

[Daccarett and Mrozowski 2002]

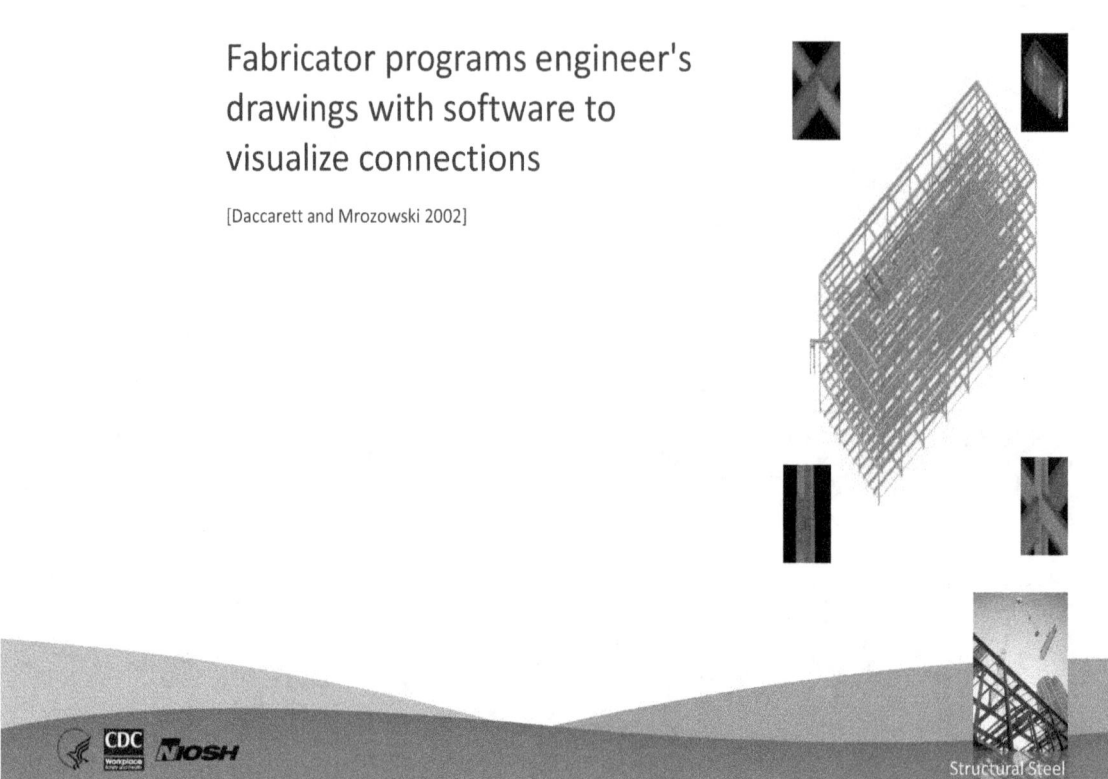

Structural Steel

NOTES

Steel detailing involves identifying the specific dimensions of every aspect of a structural steel member and connection, including bolt holes, cut-outs, and web stiffeners. Detailing is a critical step between the design and fabrication that typically involves tolerances of tenths of an inch. A small mistake in just one dimension can result in thousands of dollars of additional labor. Although some detailing is done manually, many detailers today create elaborate computer models of a structural steel frame and connections with use of sophisticated 3D detailing software. These models can then be passed on to the fabricator to allow efficient, numerically controlled, computer fabrication.

SOURCE

Daccarett V, Mrozowski T [2002]. Structural steel construction process: technical [PowerPoint]. Chicago: American Institute of Steel Construction, AISC Digital Library [www.aisc.org/content. aspx?id=21252].

 **Shop Drawings**

While detailing, fabricator makes drawings containing specifics about how to fabricate each member

[Daccarett and Mrozowski 2002]

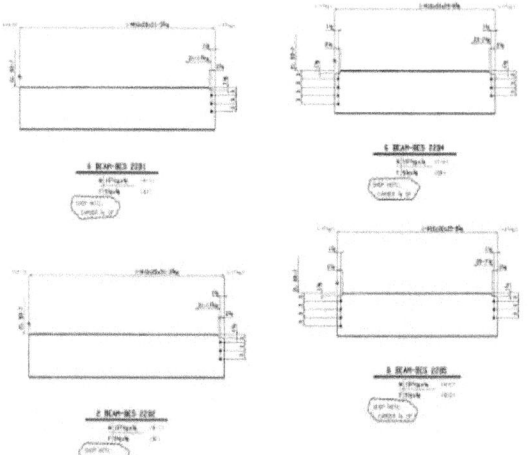

Structural Steel

NOTES

On the shop drawings the fabricator provides dimensions for machining the steel. These drawings show the length of the members, the offset from the edge of the steel face, the distance between bolt holes, and the size of those holes.

SOURCE

Daccarett V, Mrozowski T [2002]. Structural steel construction process: technical [PowerPoint]. Chicago: American Institute of Steel Construction, AISC Digital Library [www.aisc.org/content. aspx?id=21252].

 **Fabrication**

To achieve its final configuration, the steel may be

- Cut
- Sheared
- Punched
- Drilled
- Fit
- Welded

Photo courtesy of Thinkstock

Each final member is labeled with a piece mark, length, and job number for identification.

[Daccarett and Mrozowski 2002]

Structural Steel

NOTES

After the fabricator has manufactured or purchased the specific structural shape, the piece is finalized for erection in the fabrication shop. The steel may be cut, sheared, punched, drilled, fit, and welded so that it has the proper length, holes, and connection plates. Some fabrication shops are better equipped for certain types of connections. In-house detailers should design those connections to save the owner time and money. Finally, each piece is labeled with a piece mark, length, and job number so it can be identified and properly installed.

SOURCES

Daccarett V, Mrozowski T [2002]. Structural steel construction process: technical [PowerPoint]. Chicago: American Institute of Steel Construction, AISC Digital Library [www.aisc.org/content. aspx?id=21252].

Photo courtesy of Thinkstock

 **Transportation**

Members are
transported via

- Flatbed truck
- Train
- Waterways

Photo courtesy Thinkstock

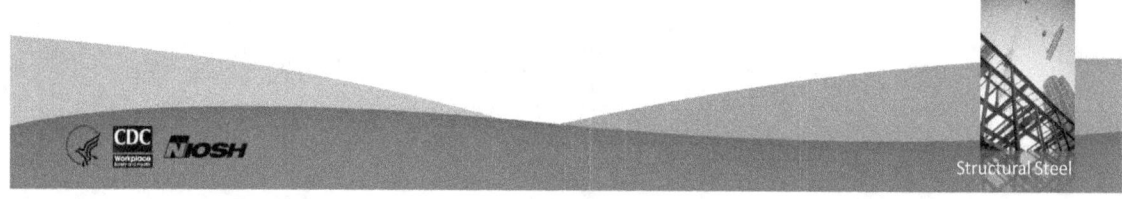

NOTES

Transportation is one of the limiting factors on beam/column length. After fabrication, members are transported by flatbed truck, waterways, railroad, or probably a combination of the three. Steel manufacturers often are located near a waterway or railway for transportation purposes.

SOURCE

Photo courtesy of Thinkstock

Erection Process

STRUCTURAL STEEL DESIGN

Erection Process

NOTES

This section summarizes the typical structural steel erection process. Many of the text and photos have been taken from the excellent instructor aids that the AISC makes available to instructors on its Web site.

Unloading and Shake-out

- Steel members are unloaded and placed on blocking to allow space for chokers to be easily attached.

- Shake-out: members are sorted on the ground to allow for efficient erection.

[Daccarett and Mrozowski 2002]

Photo courtesy of Thinkstock

NOTES

The members will most likely arrive at the site on a truck. Depending on the job site location, the members can either be unloaded and stored at the beginning of construction or, on a crowded urban site, be delivered as needed and immediately erected. When members are stored on site, they are placed on steel blocking to allow the crane chokers to be easily attached. The members also are "shaken out," or sorted on the ground to be arranged in an order that is beneficial for erection efficiency.

SOURCES

Daccarett V, Mrozowski T [2002]. Structural steel construction process: technical [PowerPoint]. Chicago: American Institute of Steel Construction, AISC Digital Library [www.aisc.org/content. aspx?id=21252].

Photo courtesy of Thinkstock

Picking and Hoisting

- Cranes lift members into place

- Hole at end of each column

- After a choker is tied around the center of gravity, multiple beams can be lifted at once

[Daccarett and Mrozowski 2002]

Photo courtesy of Thinkstock

Structural Steel

NOTES

A crane is used to lift the members into place. A column is picked up by the hole at its end and then placed on a base plate secured in the ground. Beams have chokers tied around their center of gravity. Oftentimes there will be more than one choker supporting a beam, especially during initial bolting. Multiple beams can be lifted at the same time, as shown, to lessen the time required for the crane to be moving back and forth to pick up the members.

SOURCES

Daccarett V, Mrozowski T [2002]. Structural steel construction process: technical [PowerPoint]. Chicago: American Institute of Steel Construction, AISC Digital Library [www.aisc.org/content.aspx?id=21252].

Photo courtesy of Thinkstock

Positioning and Initial Bolting

- Each beam is lowered into place, and a worker lines it up correctly with drift pins. At least two bolts are attached before the crane releases the load.

    – OSHA requirement

[Daccarett and Mrozowski 2002]

Photo courtesy of Daccarett and Mrozowski

Structural Steel

NOTES

As the crane lowers the beam into place, a construction worker lines up the connection, using drift pins. After the holes are correctly aligned, at least two bolts are installed and tightened with a spud wrench. They are not at full tightness, as this stage is only temporary. After these bolts are installed the crane may release the load. This can be very dangerous work because the beam is not secured.

SOURCE

Daccarett V, Mrozowski T [2002]. Structural steel construction process: technical [PowerPoint]. Chicago: American Institute of Steel Construction, AISC Digital Library [www.aisc.org/content. aspx?id=21252].

Final Bolting

- Once everything is in the correct position, the final bolting is performed with a torque wrench or similar tool.

[Daccarett and Mrozowski 2002]

Photo courtesy of Daccarett and Mrozowski

CDC NIOSH

Structural Steel

NOTES

Periodically, the structure is "plumbed up" to ensure it is within acceptable vertical and horizontal alignment. Once the structure is within specification, final bolting occurs. A torque wrench or other device applies a specific amount of tension to tighten the bolts.

SOURCE

Daccarett V, Mrozowski T [2002]. Structural steel construction process: technical [PowerPoint]. Chicago: American Institute of Steel Construction, AISC Digital Library [www.aisc.org/content.aspx?id=21252].

Examples of Prevention through Design

STRUCTURAL STEEL DESIGN

Examples of Prevention through Design

NOTES

This section provides a number of examples of how PtD can be applied to structural steel design. Many of these examples were taken either from a research report published by the Construction Industry Institute or from the Detailing Guide For Erector's Safety & Efficiency, published jointly by the National Institute of Steel Detailing and the Steel Erectors Association of America.

Topics

Topics	Slide numbers
Prefabrication	44–45
Access Help	46
Columns	47–50
Beams	51–54
Connections	55–67
Miscellaneous	68–77

NOTES

PtD examples and guidelines are briefly covered for each of the structural steel components shown on this slide.

Prefabrication

- Shop work is often faster than field work.

- Shop work is less expensive than field work.

- Shop work is more consistent because of the controlled environment.

- Shop work yields better quality than field work.

- With prefabrication, less work is done at high elevations, which reduces the risks of falls and falling objects.

[Toole and Gambatese 2008]

NOTES

One of the easiest ways to protect workers is to reduce the number of field connections by increasing the amount of shop fabrication and preassembly. Prefabrication and preassembly allow more of the work to be done at grade level rather than at dangerous heights. In general, fabrication in controlled environments is of better quality and can be performed more quickly. Shop fabrication can be done on a table with machines and with automated equipment. Prefabrication has other benefits: quality, time efficiency, and consistency. Prefabrication is limited by the ability to transport and install preassembled pieces.

SOURCE

Toole TM, Gambatese J [2008]. The trajectories of prevention through design in construction. J Safety Res 30(2):225–230.

 **Example: Prefabricated Truss**

- Fewer connections to make in the air

- Safer and faster

Photo courtesy of Thinkstock

Structural Steel

NOTES

Roof trusses are one example of prefabricated parts that greatly decrease the number of connections made in the air. This is safer for construction workers and allows the project to be completed faster. Other examples include prefabricated walls, bridge segments, and steel stairs.

SOURCE

Photo courtesy of Thinkstock

Access Help

- Shop-installed vertical ladders

- Bolts on ladders and platforms can be removed later or kept for maintenance

Photo courtesy of Thinkstock

Structural Steel

NOTES

Access ladders and platforms can come shop-installed or bolted on and can be used for maintenance or removed. The first way access ladders and platforms can help is by providing access to a connection that would otherwise be out of reach. This will improve the quality of work. Platforms reduce the risk of falling. Access platforms and ladders provide easier access to work, thereby reducing the number of musculoskeletal injuries caused by working in an awkward position. Shown here are an access ladder and platform on a bridge pier.

SOURCE

Photo courtesy of Thinkstock

Column Safety

- Column splices
- Tabs/Holes for safety lines
- Base plates

Photo courtesy of Thinkstock

Structural Steel

NOTES

Heavy, tall columns pose many dangers to workers. Three problem areas will be addressed here: column length, fall protection, and base plates. First, tall buildings require very long columns. For both structural and transportation reasons, they cannot be made as one huge column. Multiple columns are spliced together to reach great heights. Shown is a column splice located above a floor connection. Angles provide a seat for the beam while it is still attached to the choker during initial placement and bolting.

How can fall protection be implemented on columns? Shop-fabricated holes can be used to string safety lines between columns at proscribed heights. Shown is a safety line tab.

What kind of loads must base plates resist? The column-to-base-plate connection is a moment connection.

SOURCE

Photo courtesy of Thinkstock

Column Splices

- Have column splice around 4 feet above the working floor

  - OSHA requirement

Photo courtesy of Bucknell University facilities

Structural Steel

NOTES

Column splices are necessary for making very tall structures, because one very long column is neither transportable nor structurally desirable. A best practice is to place the column splices about 4 feet above the working floor, about chest-high for most workers. This eliminates awkward positioning such as bending over or reaching up to work. Placing the splices here also prevents column connections from sharing the same location as the beam connections and is an OSHA requirement. Incidentally, the safety line runs through holes placed 39 inches to 42 inches on the column above the working floor. Why are column splices placed every two stories?

SOURCES

OSHA steel erection e-tool: structural stability. Washington, DC: Occupational Safety and Health Administration [www.osha.gov/SLTC/etools/steelerection/structural.html].

Photo courtesy of Bucknell University facilities

Holes for Safety Lines

- Include holes at 21 inches and 42 inches for guardrails

- Additional, higher holes can also be included for lifeline support

[Gambatese 1996; NISD and SEAA 2009]

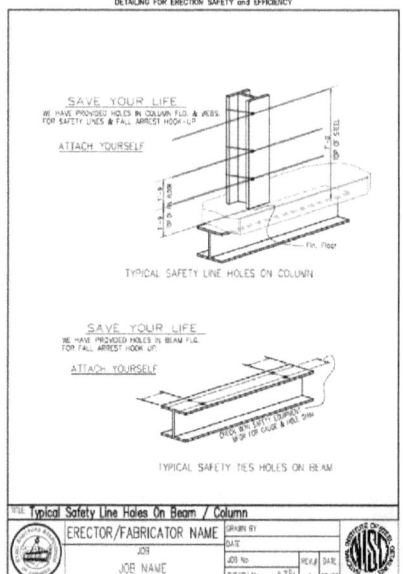

Structural Steel

NOTES

Safety lines are necessary for fall protection. There are a few ways to attach to a fall protection system. Steel cables can be strung between holes in the columns, and the worker can tie off to the cable. This method has the benefit of mobility. The worker can walk around and the tie off point follows along the cable. Often there are tabs for installing safety handrails.

The designer should specify hole spacing on the structural drawings with a note to drill holes in the shop. For the guardrails, include holes at 21 inches and 42 inches from the elevation of the floor. These heights should prevent workers from getting under the rail or over the rail. The installation of tie-off points on the steel cables is facilitated by placing holes higher up (shown in the picture at 7 feet), making it overhead.

SOURCES

Gambatese JA [1996]. Design for safety. RS101-1. Austin, TX: Construction Industry Institute.

National Institute of Steel Detailing and Steel Erectors Association of America [2009]. Detailing guide for erector's safety & efficiency. Volume II [www.nisd.org and www.seaa.net]. Oakland, CA, and Greensboro, NC: National Institute of Steel Detailing and Steel Erectors Association of America.

Base Plates

- Column **base plates** should always have at least 4 anchor rods bolted in

  – OSHA Requirement

[Gambatese 1996; NISD and SEAA 2009; OSHA 29 CFR 1926-755]

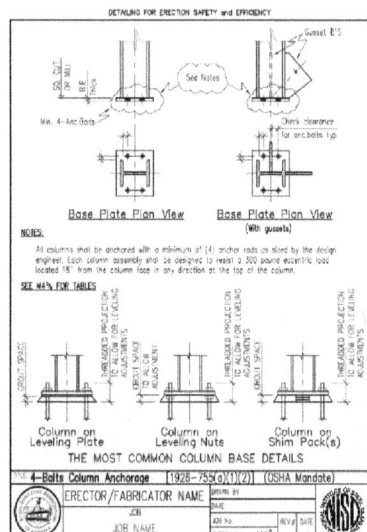

NOTES

When the columns have no beams attached to them, they are cantilevered. The loads from the stories above create significant moments at the base. Proper base plate practices are crucial to construction safety. OSHA requires that there be at least four anchor rods bolted into the base plate. Base plates must resist a 300-pound eccentric load applied to the column face at the top of the column during construction, per the same OSHA requirement. Why is this required? This will first and foremost prevent the column from falling over or "rolling over." It will also stabilize the column.

SOURCES

Gambatese JA [1996]. Design for safety. RS101-1. Austin, TX: Construction Industry Institute.

National Institute of Steel Detailing and Steel Erectors Association of America [2009]. Detailing guide for erector's safety & efficiency. Volume II [www.nisd.org and www.seaa.net]. Oakland, CA, and Greensboro, NC: National Institute of Steel Detailing and Steel Erectors Association of America.

OSHA. Regulations Part 1926–Safety and Health Regulations for Construction, Subpart R: Steel Erection, Column Anchorage. 29 CFR 1926.755. Washington, DC: U.S. Department of Labor, Occupational Safety and Health Administration.

Beams and Girders

Workers walk on beams to get to connections or other columns, a common fall hazard. Increase safety by considering

- Beam width

- Use of cantilevers

- Ability to support lifelines

Photo courtesy of Thinkstock

Structural Steel

NOTES

Beams and girders not only transfer floor loads to the columns, they often serve as a transportation route on a structure. Workers walk across them to get from place to place. Making this activity safer, whether though lifelines, flange width, or general design choice, should be a priority.

SOURCE

Photo courtesy of Thinkstock

 **Beam Width**

- For walking safely, use beams with a minimum beam width of 6 inches.

[Gambatese 1996]

Photo courtesy of Thinkstock

Structural Steel

NOTES

Most beams have a taller cross-section than flange width to increase the moment of inertia about the x-axis. Why would you specify beams with a flange width that is six (6) inches wide or wider?

SOURCES

Gambatese JA [1996]. Design for safety. RS101-1. Austin, TX: Construction Industry Institute.

Photo courtesy of Thinkstock

 **Use of Cantilevers**

Minimize the use of cantilevers, which

- are not good for tying off

- pose connection difficulties

[Gambatese 1996]

Photo courtesy of Thinkstock

Structural Steel

NOTES

Cantilevers are subject to several safety issues. Look at the photograph, and look at the set of cantilevers jutting out. Now imagine that columns on the upper story are not there yet. Where can workers tie off? How safe it that?

Cantilevers are connected on one end. During installation, how would you design the connection to resist the moment due to the weight of the member? You may need more than two preliminary bolts to resist the moment. How would you connect the upper tier column to the end of the cantilevered beam?

SOURCES

Gambatese JA [1996]. Design for safety. RS101-1. Austin, TX: Construction Industry Institute.

Photo courtesy of Thinkstock

 **Ability to Support Lifelines**

- Design beams near or above openings to be able to support lifelines

- Contract drawings should make clear how many lifelines each beam can support, and at what locations they can be attached

[Gambatese 1996; NISD and SEAA 2009; OSHA 29 CFR 1926.502(d)(15)]

Photo courtesy of Thinkstock

NOTES

How many people can tie off to the same anchor point?

Should that be indicated on the structural drawings?

How likely is it that you could overload the anchor point?

Identifying and/or designing anchor points for fall protection systems such as body harnesses and lanyards is an example of how designers can use their understanding of structural engineering principles to protect workers, during the construction and future maintenance. Specifically, designers can place floor perimeter beams and beams above floor openings to support lanyards, place lanyard connection points along the beams, and note on the contract drawings which beams are designed to support lanyards, how many lanyards, and at what locations along the beams.

The idea of identifying anchorage points on construction drawings is in accordance with Appendix C to Subpart M (Fall Protection) of the federal OSHA Standards for Construction, though it is listed as a nonmandatory practice:

(h) Tie-off considerations (1) "One of the most important aspects of personal fall protection systems is fully planning the system before it is put into use. Probably the most overlooked component is planning for suitable anchorage points. Such planning should ideally be done before the structure or building is constructed so that anchorage points can be incorporated during construction for use later for window cleaning or other building maintenance. If properly planned, these anchorage points may be used during construction, as well as afterwards."

SOURCES

Gambatese JA [1996]. Design for safety. RS101-1. Austin, TX: Construction Industry Institute.

National Institute of Steel Detailing and Steel Erectors Association of America [2009]. Detailing guide for erector's safety & efficiency. Volume II [www.nisd.org and www.seaa.net]. Oakland, CA, and Greensboro, NC: National Institute of Steel Detailing and Steel Erectors Association of America.

Occupational Safety and Health Administration (OSHA). Regulations Part 1926–Safety and Health Regulations for Construction, Subpart M–Fall Safety, Standard Number 1926 Subpart M App C–Personal Fall Arrest Systems. Non-Mandatory Guidelines for Complying with 1926.502(d). Washington, DC: U.S. Department of Labor, OSHA. 29 CFR 1926.502(d).

Photo courtesy of Thinkstock

 **Connections**

Connections are very important but can be very difficult to install. There are two main tools for making connections:

- Bolts
- Welds

Photos courtesy of AISC

Structural Steel

NOTES

Faulty connections can cause fatal accidents!

The two primary ways of making connections are bolting and welding.

SOURCES

American Institute of Steel Construction [2004]. Bolting and welding [PowerPoint]. AISC Digital Library. Chicago: American Institute of Steel Construction, AISC Digital Library [www.aisc.org/content.aspx?id=21760].

Photo courtesy of AISC

Bolts

For safe bolted connections, consider:

- Self-supporting connections

- Double connections

- Erection aid: "dummy holes"

- Bolt sizes

- Minimum number of bolts

- Awkward or dangerous connection locations

Photo courtesy of AISC

Structural Steel

NOTES

Bolts are the most common method of making connections because they are cheaper and faster than welds. Like all connections, mistakes in designing or in the process of bolting can be very costly. The next few slides present specific examples of how careful design can increase safety.

SOURCES

American Institute of Steel Construction [2004]. Bolting and welding [PowerPoint]. AISC Digital Library. Chicago: American Institute of Steel Construction, AISC Digital Library [www.aisc.org/content.aspx?id=21760].

Photo courtesy of AISC

Self-Supporting Connections

- Avoid hanging connections
- Consider using beam seats

[Gambatese 1996; NISD and SEAA 2009]

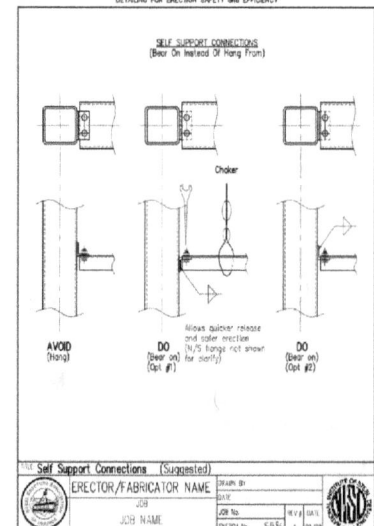

NOTES

Hanging connections are those made in which the member "hangs" from a tab, such as in the left drawing. The only thing supporting the member while the connections are fastened is the choker(s) from the crane. Self-supporting connections are made while the member rests on some other piece. As can be seen in the second and third drawings, the member is resting on an angle. A beam seat is a shelf that is attached to a column to support beams while the member is connected to the structure. This extra support, which makes it much more stable, is easier and safer to connect.

SOURCES

Gambatese JA [1996]. Design for safety. RS101-1. Austin, TX: Construction Industry Institute.

National Institute of Steel Detailing and Steel Erectors Association of America [2009]. Detailing guide for erector's safety & efficiency. Volume II [www.nisd.org and www.seaa.net]. Oakland, CA, and Greensboro, NC: National Institute of Steel Detailing and Steel Erectors Association of America.

 **Double Connections**

- Avoid beams of common depth connecting into the column web at the same location.

- If double connections are necessary, design them to have full support during the connection process.

    − OSHA requirement

[Gambatese 1996; NISD and SEAA 2009; OSHA 29 CFR 1926-756]

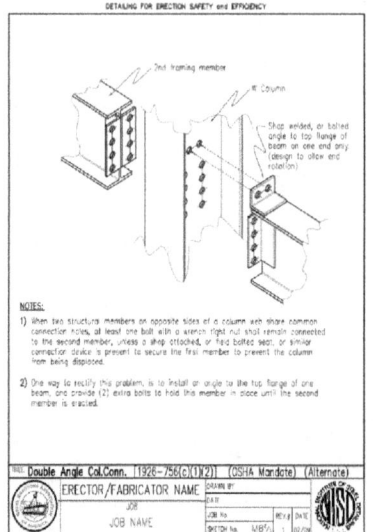

NOTES

Double connections are a major danger on the construction site, and they are most commonly caused by beams of common depth connecting into the column web at the same location (which is why you should avoid that if possible). The problem with double connections is that they share a common bolt, so until the bolt is secured on the threaded side, it can just be pushed out. However, should one absolutely need to make a double connection, there are ways to do it safer, and each way involves securing one member without affecting the other. The first way, shown in the picture, is to attach an angle to one member only and secure the bolts on that angle first. This way, even if bolts get pushed out, the member is still supported by the angle. A second way of accomplishing this is to put an extra row of holes on one side and not the other, as seen in the next slide. This way, even if the bolts get pushed out, the member is still supported by the extra row.

SOURCES

Gambatese JA [1996]. Design for safety. RS101-1. Austin, TX: Construction Industry Institute.

National Institute of Steel Detailing and Steel Erectors Association of America [2009]. Detailing guide for erector's safety & efficiency. Volume II [www.nisd.org and www.seaa.net]. Oakland, CA, and Greensboro, NC: National Institute of Steel Detailing and Steel Erectors Association of America.

OSHA. Regulations Part 1926–Safety and Health Regulations for Construction, Subpart R: Steel Erection, Beams and Columns. 29 CFR 1926.756. Washington, DC: U.S. Department of Labor, Occupational Safety and Health Administration.

Alternate Double Connection

Photos courtesy of AISC

Structural Steel

NOTES

The connection has four common bolts and two that are for only one beam.

SOURCES

American Institute of Steel Construction [2004]. Bolting and welding [PowerPoint]. AISC Digital Library. Chicago: American Institute of Steel Construction, AISC Digital Library [www.aisc.org/content.aspx?id=21760].

Photo courtesy of AISC

Erection Aids: "Dummy Holes"

- Provide an extra "dummy hole" in the connection, where a spud wrench can be inserted

- This is most appropriate when there are only two bolts

[Gambatese 1996]

Photo courtesy Bucknell University facilities

Structural Steel

NOTES

A "dummy hole" is for erection purposes only. It allows a spud wrench to be inserted, keeping the member in place while the bolts are being installed. If there are only two bolts in a connection it is safer to use the dummy hole so that the spud wrench remains while both bolts are fastened. When do you specify a dummy hole? When do you pull out the spud wrench?

SOURCES

Gambatese JA [1996]. Design for safety. RS101-1. Austin, TX: Construction Industry Institute.

Photo courtesy of Bucknell University facilities

Bolt Sizes

- Use as few bolt sizes as possible

[Gambatese 1996]

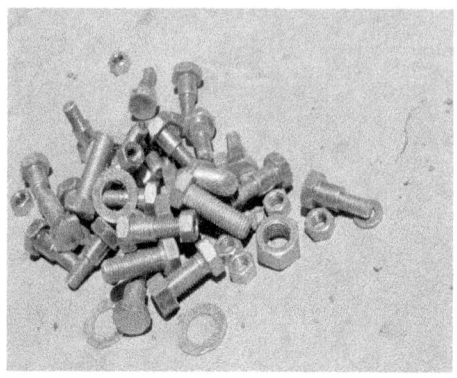

Photo courtesy of Thinkstock

Structural Steel

NOTES

Are the bolts in the picture all the same size?

Does it matter?

Tracking several sizes of bolts to ensure proper installation costs more than using one size of bolt for all connections.

SOURCES

Gambatese JA [1996]. Design for safety. RS101-1. Austin, TX: Construction Industry Institute.

Photo courtesy of Thinkstock

Minimum Number of Bolts

- Use a minimum of two bolts per connection
  - OSHA requirement

[Gambatese 1996]

Photo courtesy of AISC

Structural Steel

NOTES

Why are two bolts safer than one? Prevent rotation by using two or more bolts for each connection. Whereas a single bolt may allow for some rotation of the member, a two-bolt connection will not. If one bolt fails, the member may shake but will not fall. Per OSHA, at no time may a member be supported by a connection with only one bolt.

SOURCES

American Institute of Steel Construction [2004]. Bolting and welding [PowerPoint]. AISC Digital Library [www.aisc.org/content.aspx?id=21760].

Gambatese JA [1996]. Design for safety. RS101-1. Austin, TX: Construction Industry Institute.

Photo courtesy of AISC

 **Immediate Stability**

Provide pin-holed or bolted
connections to provide
immediate stability after
placement of members

[Gambatese 1996]

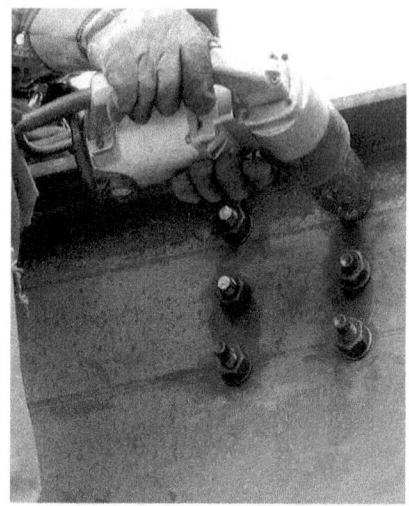

Photo courtesy of AISC

NOTES

In the field, bolts are preferable to welds. Bolted connections are stable as soon as the first two
bolts are in, and they are secure as soon as the final bolting is done. Welds require more time and
multiple passes. During this time there is the danger of a piece getting moved, requiring the weld
to be redone. Welds also require specific weather and temperature conditions. Construction could
be delayed if they are used. If a weld fails, the member may fall. Using shop welds prevents these
problems because the welding is done in a much better, controlled environment and the piece is
ready for installation. The preferable connection involves a shop-welded plate on a member to be
field-bolted.

SOURCES

American Institute of Steel Construction [2004]. Bolting and welding [PowerPoint]. AISC Digital
Library [www.aisc.org/content.aspx?id=21760].

Gambatese JA [1996]. Design for safety. RS101-1. Austin, TX: Construction Industry Institute.

Photo courtesy of AISC

Avoid Awkward or Dangerous Connection Locations

- Time-consuming and dangerous

- Can cause strain

[Gambatese 1996; NISD and SEAA 2009]

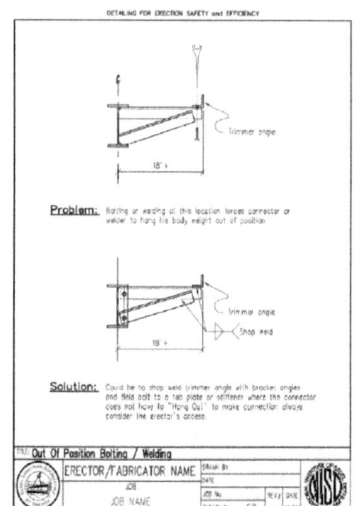

NOTES

Attempting to bolt at awkward angles is both time-consuming and dangerous. Overhead connections require workers to strain to reach them; in the top diagram, the only way to get to the connection is to put a lift underneath it and work overhead. This can easily result in injuries to the back. Such situations can be avoided simply by thinking about how to better position the connection that will be made. In the example shown, the field connection has been moved from the overhang to the base plate, where the bolts can be installed at chest level and without the need for any stretching.

SOURCES

Gambatese JA [1996]. Design for safety. RS101-1. Austin, TX: Construction Industry Institute.

National Institute of Steel Detailing and Steel Erectors Association of America [2009]. Detailing guide for erector's safety & efficiency. Volume II [www.nisd.org and www.seaa.net]. Oakland, CA, and Greensboro, NC: National Institute of Steel Detailing and Steel Erectors Association of America.

 **Welds**

For safer welded connections:

- Avoid awkward or dangerous connection locations
- Immediate stability
- Welding location
- Welding material

Photo courtesy of Thinkstock

NOTES

Welding makes a connection by fusing the base metals with a filler material. Welders must go through extensive training to be able to weld in any situation. Arc welds are common in the field and use a current through the base metals to melt the material. Metal inert gas (MIG) welding uses a gas shield and a continuously fed filler wire to make a weld. This is generally cheaper, better, and faster but is typically done in the shop.

Tungsten inert gas (TIG) welding uses a constant current through a nonconsumable tungsten electrode to join stainless steel to a non-ferrous metal. The weld area is shielded by an inert gas such as argon.

SOURCE

Photo courtesy of Thinkstock

 **Welding Locations**

- Specify shop welding rather than field welding

- If field welds are necessary, design them in convenient locations

[Gambatese 1996]

Photo courtesy of Thinkstock

Structural Steel

NOTES

Welding positions (flat, horizontal, vertical, or overhead) vary in their level of difficulty. Flat welds are done on top of a surface and are the easiest to perform. Horizontal and vertical welds are on the sides of a member and are more difficult because there is no support underneath. Overhead welds are the most difficult and dangerous from an ergonomic standpoint and should be avoided when possible.

SOURCES

Gambatese JA [1996]. Design for safety. RS101-1. Austin, TX: Construction Industry Institute.

Photo courtesy of Thinkstock

Welding Material

Welding can be a fire hazard and can emit toxic fumes. Always be aware of what material is being welded.

[Gambatese 1996; Sperko Engineering Services 1999]

Photo courtesy of Thinkstock

Structural Steel

NOTES

Welding requires very high temperatures to join two base metals. Welding rods are covered by a mixture called flux, which creates a brittle cover called slag over the finished weld. Dropping slag can burn workers and may become a fire hazard. Material finishes should be considered when choosing either a bolted or welded connection. Galvanized steel includes zinc, which vaporizes during welding. Exposure to metal vapor containing non-ferrous metals (such as chromium, cadmium, and silver) and metal oxides (such as zinc oxide and magnesium oxide) can result in metal fume fever. Symptoms include fever, chills, headache, thirst, and nausea and can last up to 2 days.

SOURCES

Gambatese JA [1996]. Design for safety. RS101-1. Austin, TX: Construction Industry Institute.

Sperko Engineering Services [1999]. Welding galvanized steel—safely. Greensboro, NC: Sperko Engineering Services [www.sperkoengineering.com/html/articles/WeldingGalvanized.pdf].

Photo courtesy of Thinkstock

 **Other Methods for Safer Construction**

Address these factors:

- Sharp corners

- Access problems

- Temporary bracing

- Crane safety

- Member placement

- Tripping hazards

Photo courtesy of Thinkstock

Structural Steel

NOTES

The following slides present miscellaneous ways to help prevent construction difficulties through design.

SOURCE

Photo courtesy of Thinkstock

Avoid Sharp Corners

- Corners can cause clothing or wires to snag, resulting in falling objects or tripping hazards

- Corners can cause scratches or cuts

[NISD and SEAA 2009]

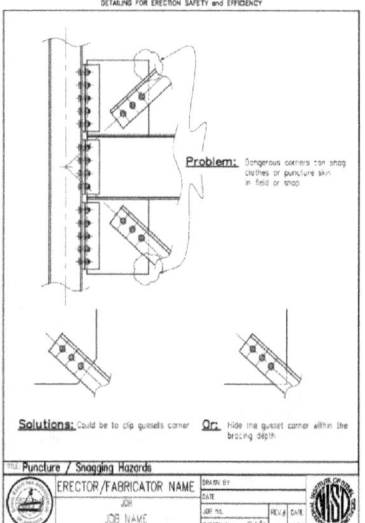

NOTES

Sharp corners on plates can snag clothing or wires and can result in falling objects or tripping hazards. A sharp corner can cause scratches or cuts. Two simple solutions are to cut off the sharp corner or to cover the corner with the bracing.

SOURCE

National Institute of Steel Detailing and Steel Erectors Association of America [2009]. Detailing guide for erector's safety & efficiency. Volume II [www.nisd.org and www.seaa.net]. Oakland, CA, and Greensboro, NC: National Institute of Steel Detailing and Steel Erectors Association of America.

Access Problems

Complicated connections
take time to complete
and are dangerous if they
require awkward positioning,
so consider

- Adequacy of space for
  making connections

- Small column size access

- Hand trap danger

Photo courtesy of Thinkstock

Structural Steel

NOTES

Connections can become very crowded and complicated. Sometimes the design makes
construction difficult. Here are some ways to simplify connections.

SOURCE

Photo courtesy of Thinkstock

Provide Enough Space for Connections

- There may not be enough space for common tools

- These connections can be made better by clipping away portions or increasing distances

[NISD and SEAA 2009]

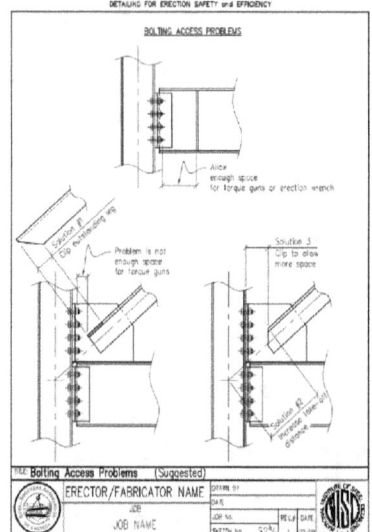

NOTES

A connection with many bolts and members is difficult to accomplish. Every piece installed means less space for workers to maneuver. There are two ways to improve constructability: clip away unnecessary portions or increase take-off distances.

SOURCE

National Institute of Steel Detailing and Steel Erectors Association of America [2009]. Detailing guide for erector's safety & efficiency. Volume II [www.nisd.org and www.seaa.net]. Oakland, CA, and Greensboro, NC: National Institute of Steel Detailing and Steel Erectors Association of America.

 **Small Column Size Access**

- Small column depth can make connections difficult

- Access to bolts can be blocked by the column flanges

- Attach a tab to the column

[NISD and SEAA 2009]

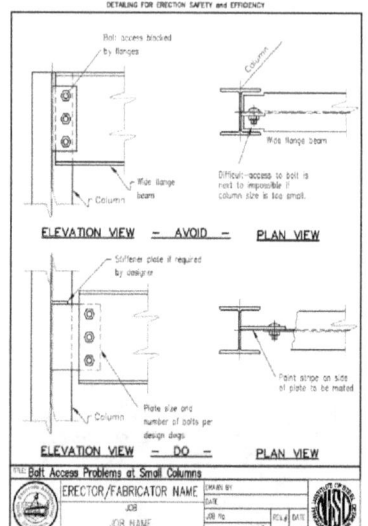

NOTES

Small columns are another challenge. The flange width may not leave enough space to access the bolts or allow a spud wrench to rotate. One solution is to attach a tab to the column. Stiffener plates may be shop-welded to the column to allow more space for the connection.

SOURCE

National Institute of Steel Detailing and Steel Erectors Association of America [2009]. Detailing guide for erector's safety & efficiency. Volume II [www.nisd.org and www.seaa.net]. Oakland, CA, and Greensboro, NC: National Institute of Steel Detailing and Steel Erectors Association of America.

Hand Trap

- The situation shown can create a dangerous hand trap

- A solution is to cut out a section of the flange to allow access to the bolts

[NISD and SEAA 2009]

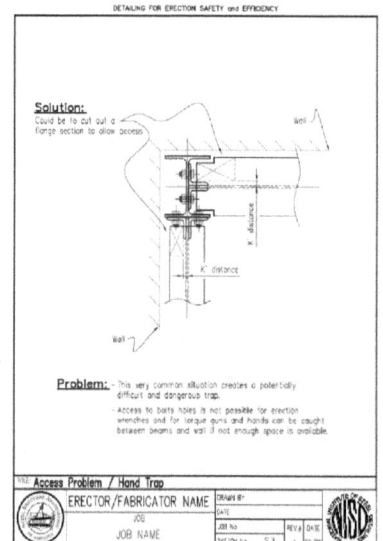

NOTES

A connection made at the intersection of two existing walls can create a dangerous hand trap. Tools may not fit in the available space between the wall and column. One solution is to cut out a section of the flange to allow better access to the bolts. This type of modification is typically done in the shop. Shop drawings should be reviewed by a structural engineer.

SOURCE

National Institute of Steel Detailing and Steel Erectors Association of America [2009]. Detailing guide for erector's safety & efficiency. Volume II [www.nisd.org and www.seaa.net]. Oakland, CA, and Greensboro, NC: National Institute of Steel Detailing and Steel Erectors Association of America.

Know Approximate Sizes of Tools

"Knuckle-busting" – workers' knuckles get damaged from trying to fit their hands into a tight space

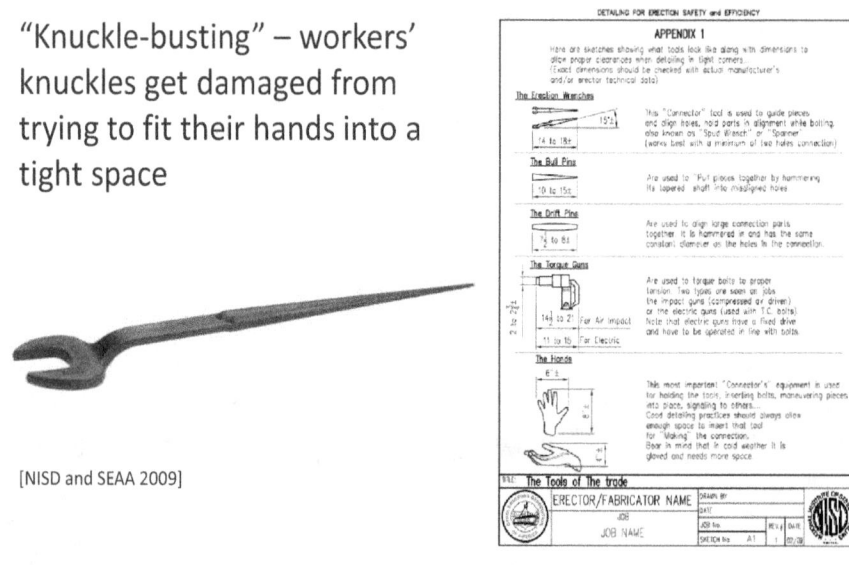

[NISD and SEAA 2009]

Structural Steel

NOTES

Making the cramped connections can result in "knuckle busting," in which workers' hands rub against connection pieces. Consider tool size when designing connections.

SOURCE

National Institute of Steel Detailing and Steel Erectors Association of America [2009]. Detailing guide for erector's safety & efficiency. Volume II [www.nisd.org and www.seaa.net]. Oakland, CA, and Greensboro, NC: National Institute of Steel Detailing and Steel Erectors Association of America.

Cranes and Derricks

- Erection and disassembly must be carefully planned.

- Site layout affects crane maneuverability.

- Show site utilities on plans.

- Comply with OSHA standards.

Photo courtesy of Walter Heckel

OSHA comprehensive crane standard: www.osha.gov/FedReg_osha_pdf/FED20100809.pdf.
Regulation text: www.osha.gov/cranes-derricks/index.html.

NOTES

Cranes are used to lift steel members into place. Cranes are the most complex machines on a construction site. Crane erection and disassembly must be carefully planned. Where do you place the crane? Ideally, the crane can lift all members from one location without interfering with any other operations. The biggest danger in site layout is overhead power lines. Although it is the contractor's responsibility to deal with power lines, the designer can help by including the power line locations on the plans.

Another problem, overturning, is often the result of moments created by the load. Cherry pickers are particularly susceptible. Cranes operate within a range defined by the mass of the crane, the length of the boom, and the mass of the load. Operators may be tempted to extend the boom a few more feet to pick up a load, when it would be safer to move the crane closer. As the load is lifted, the crane tips.

Another hazard is boom collapse. In this instance, the lift exceeds the design limits of the boom. There is always the possibility that the operator will lose control of the load, especially when it is windy. A swinging load may impact adjacent structures or touch a power line. In several instances, the crane operator has died when the load swung back into the cab.

SOURCES

The OSHA comprehensive crane standard is available at www.osha.gov/FedReg_osha_pdf/ FED20100809.pdf. The regulation text is available at www.osha.gov/cranes-derricks/index.html.

Photo courtesy of Walter Heckel

Member Placement

- Members need sufficient space to fit between columns

- Members without enough space could cause columns to tilt

[NISD and SEAA 2009]

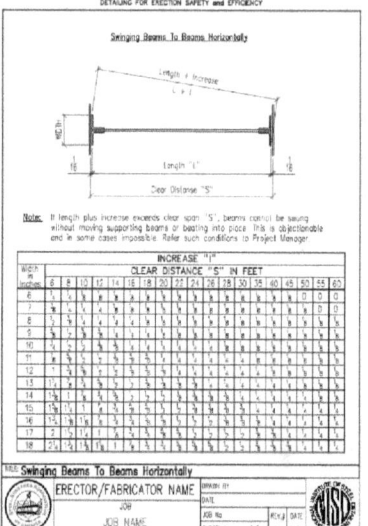

NOTES

During erection of the structure, some beams will need to be lifted and placed between two columns. Sometimes a beam will not fit unless the columns are moved, because of the width of the beam flanges. (An extreme example can be shown with a piece of paper. If you have just enough space for the piece of paper, then you cannot have one end connected and rotate the other side in.) In some cases, the column can be tilted to make the fit, which is dangerous because it could fall over. In order to ensure the beams can be easily swung in, this table is provided in the Detailing Guide. The larger the ratio of width to length, the more impact the corners of the beam will have during the attempt to swing it in. The table indicates the increase in length required to fit a beam of a certain width into a certain space.

SOURCE

National Institute of Steel Detailing and Steel Erectors Association of America [2009]. Detailing guide for erector's safety & efficiency. Volume II [www.nisd.org and www.seaa.net]. Oakland, CA, and Greensboro, NC: National Institute of Steel Detailing and Steel Erectors Association of America.

Tripping Hazards

Avoid having connections on top
of beams and joists.

[NISD and SEAA 2009; OSHA 29 CFR 1926-754]

Image courtesy of Thinkstock

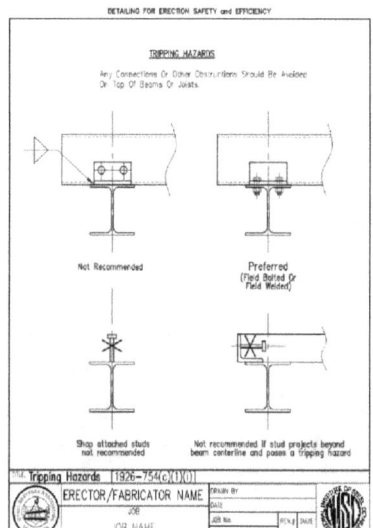

Structural Steel

NOTES

Tripping may be a minor embarrassment, or it could be fatal. The steel designer can help prevent trips by avoiding connections on the top of beams and joists. There are an infinite number of other tripping hazards, such as shear studs, random objects, or moisture from the weather.

SOURCES

National Institute of Steel Detailing and Steel Erectors Association of America [2009]. Detailing guide for erector's safety & efficiency. Volume II [www.nisd.org and www.seaa.net]. Oakland, CA, and Greensboro, NC: National Institute of Steel Detailing and Steel Erectors Association of America.

OSHA. Regulations Part 1926–Safety and Health Regulations for Construction, Subpart R: Steel Erection, Tripping Hazards. 29 CFR 1926.754(c)(1)(1). Washington, DC: U.S. Department of Labor, Occupational Safety and Health Administration.

Photo courtesy of Thinkstock

Summary

 **Recap**

- **Prevention through Design (PtD)** is an emerging process for saving lives, time, and money.

- PtD is the smart thing to do and the right thing to do.

- Although site safety is the contractor's responsibility, the designer has the ethical duty to create drawings with good constructability.

- There are tools and examples to facilitate PtD.

NOTES

- PtD saves lives, time, and money.
- PtD is the ethical thing to do.
- Good constructability is the designer's responsibility.

Help make the workplace safer...

Include *Prevention through Design* **concepts in your projects.**

For more information, please contact the National Institute for Occupational Safety and Health (NIOSH) at

> **Telephone: (513) 533–8302**
> **E-mail:** preventionthroughdesign@cdc.gov

Visit these NIOSH Prevention through Design Web sites:

www.cdc.gov/niosh/topics/PtD

www.cdc.gov/niosh/programs/PtDesign

Structural Steel

NOTES

This presentation was intended to provide examples of construction hazards and risks that could be positively or negatively affected by design decisions. It is certainly not comprehensive in any way. All members of the construction project team (owner, designers, contractors, and safety professionals) must attempt to learn more about construction site safety early in the built environment's life cycle. The earlier more is learned, the more effective and safer the process can be. Each party has a role to play. The United Kingdom and Australia have promulgated designers' roles and responsibilities for safe construction design. Designers are still learning to identify and manage risks to provide safer and healthier designs. We encourage the infusion of construction and safety knowledge into the design team and design reviews. Organizations and individuals seeking to positively impact construction workers' safety and health through design will need first an open mind and second a holistic view of what factors influence workers' actions and inactions.

Are there any questions?

SOURCES

NIOSH PtD Program Web sites:

www.cdc.gov/niosh/topics/PtD/

www.cdc.gov/niosh/programs/PtDesign/

References

American Institute of Industrial Hygienists [AIHA] [2008]. Strategy to demonstrate the value of industrial hygiene [www.aiha.org/votp_NEW/pdf/votp_exec_summary.pdf].

ANSI/AIHA [2005]. American national standard for occupational health and safety management systems. New York: American National Standards Institute, Inc. ANSI/AIHA Z10-2005.

American Institute of Steel Construction [2004]. Bolting and welding [PowerPoint]. AISC Digital Library. Chicago: American Institute of Steel Construction, AISC Digital Library [www.aisc.org/content.aspx?id=21760].

Behm M [2005]. Linking construction fatalities to the design for construction safety concept. Safety Sci 43:589–611.

Bureau of Labor Statistics [BLS] [2003–2009]. Census of fatal occupational injuries. Washington, DC: U.S. Department of Labor, Bureau of Labor Statistics [www.bls.gov/iif/oshcfoi1.htm].

BLS [2003–2009]. Current population survey. Washington, DC: U.S. Department of Labor, Bureau of Labor Statistics [www.bls.gov/cps/home.htm].

BLS [2006]. Injuries, illnesses, and fatalities in construction, 2004. By Meyer SW, Pegula SM. Washington, DC: U.S. Department of Labor, Bureau of Labor Statistics, Office of Safety, Health, and Working Conditions [www.bls.gov/opub/cwc/sh20060519ar01p1.htm].

BLS [2011a]. Census of fatal occupational injuries. Washington, DC: U.S. Department of Labor, Bureau of Labor Statistics [www.bls.gov/news.release/cfoi.t02.htm].

BLS [2011b]. Injuries, illnesses, and fatalities (IIF). Washington, DC: U.S. Department of Labor, Bureau of Labor Statistics [www.bls.gov/iif/home.htm].

CFR. Code of Federal Regulations. Washington, DC: U.S. Government Printing Office, Office of the Federal Register.

CPWR [2008]. The construction chart book. 4th ed. Silver Spring, MD: Center for Construction Research and Training.

Daccarett V, Mrozowski T [2002]. Structural steel construction process: technical [PowerPoint]. Chicago: American Institute of Steel Construction, AISC Digital Library [www.aisc.org/content.aspx?id=21252].

Driscoll TR, Harrison JE, Bradley C, Newson RS [2008]. The role of design issues in work-related fatal injury in Australia. J Safety Res 39(2):209–214.

European Foundation for the Improvement of Living and Working Conditions [1991]. From drawing board to building site (EF/88/17/FR). Dublin: European Foundation for the Improvement of Living and Working Conditions.

Gambatese JA, Hinze J, Haas CT [1997]. Tool to design for construction worker safety. J Arch Eng 3(1):2–41.

Gambatese JA [1996]. Design for safety. RS101-1. Austin, TX: Construction Industry Institute.

Hecker S, Gambatese J, Weinstein M [2005]. Designing for worker safety: moving the construction safety process upstream. Prof Saf 50(9):32–44.

Hinze J, Wiegand F [1992]. Role of designers in construction worker safety. Journal of Construction Engineering and Management 118(4):677–684.

Lipscomb HJ, Glazner JE, Bondy J, Guarini K, Lezotte D [2006]. Injuries from slips and trips in construction. Appl Ergonomics 37(3):267–274.

Main BW, Ward AC [1992]. What do engineers really know and do about safety? Implications for education, training, and practice. Mechanical Engineering 114(8):44–51.

New York State Department of Health [2007]. A plumber dies after the collapse of a trench wall. Case report 07NY033 [www.cdc.gov/niosh/face/pdfs/07NY033.pdf].

NIOSH Fatality Assessment and Control Evaluation (FACE) Program [1983]. Fatal incident summary report: scaffold collapse involving a painter. FACE 8306 [www.cdc.gov/niosh/face/In-house/full8306.html].

National Institute of Steel Detailing and Steel Erectors Association of America [2009]. Detailing guide for erector's safety & efficiency. Volume II. [www.nisd.org and www.seaa.net]. Oakland, CA, and Greensboro, NC: National Institute of Steel Detailing and Steel Erectors Association of America.

NOHSC [2001]. CHAIR safety in design tool. New South Wales, Australia: National Occupational Health & Safety Commission.

OSHA [2001]. Standard number 1926.760: fall protection. Washington, DC: U.S. Department of Labor, Occupational Safety and Health Administration.

OSHA [ND]. Fatal facts accident reports index [foreman electrocuted]. Accident summary no. 17 [www.setonresourcecenter.com/MSDS_Hazcom/FatalFacts/Index.htm].

OSHA [ND]. Fatal facts accident reports index [laborer struck by falling wall]. Accident summary no. 59 [www.setonresourcecenter.com/MSDS_Hazcom/FatalFacts/Index.htm].

OSHA [ND]. OSHA steel erection e-tool: structural stability. Washington, DC: Occupational Safety and Health Administration [www.osha.gov/SLTC/etools/steelerection/structural.html].

OSHA [ND]. Regulations Part 1926 –Safety and Health Regulations for Construction, Subpart M—Fall Safety, Standard Number 1926 Subpart M App C –Personal Fall Arrest Systems. Non-Mandatory Guidelines for Complying with 1926.502(d). Washington, DC: U.S. Department of Labor, OSHA.

OSHA. Regulations Part 1926–Safety and Health Regulations for Construction, Subpart R: Steel Erection, Tripping Hazards. 29 CFR 1926.754(c)(1)(1). Washington, DC: U.S. Department of Labor, Occupational Safety and Health Administration.

OSHA. Regulations Part 1926–Safety and Health Regulations for Construction, Subpart R: Steel Erection, Column Anchorage. 29 CFR 1926.755. Washington, DC: U.S. Department of Labor, Occupational Safety and Health Administration.

OSHA. Regulations Part 1926–Safety and Health Regulations for Construction, Subpart R: Steel Erection, Beams and Columns. 29 CFR 1926.756. Washington, DC: U.S. Department of Labor, Occupational Safety and Health Administration.

Sperko Engineering Services [1999]. Welding galvanized steel—safely. Greensboro, NC: Sperko Engineering Services [www.sperkoengineering.com/html/articles/WeldingGalvanized.pdf].

Szymberski R [1997]. Construction project planning. TAPPI J *80*(11):69–74.

Toole TM [2005]. Increasing engineers' role in construction safety: opportunities and barriers. Journal of Professional Issues in Engineering Education and Practice *131*(3):199–207.

Toole TM, Gambatese J [2008]. The trajectories of prevention through design in construction. J Safety Res *30*(2):225–230.

USC. United States Code. Washington, DC: U.S. Government Printing Office.

Other Sources

American Society of Civil Engineers [ASCE] [www.asce.org/Content.aspx?id=7231]

NIOSH Fatality Assessment and Control Evaluation Program [www.cdc.gov/niosh/face/]

National Society of Professional Engineers [NSPE] [www.nspe.org/ethics]

NIOSH PtD Web sites:
  www.cdc.gov/niosh/topics/PtD/
  www.cdc.gov/niosh/programs/PtDesign/

OSHA Fatal Facts [www.setonresourcecenter.com/MSDS_Hazcom/FatalFacts/index.htm]

OSHA home page [www.osha.gov/pls/oshaweb/owastand.display_standard_group?p_toc_level=1&p_part_number=1926]

OSHA Anchorage Standard 29 CFR 1926.502(d)(15)

OSHA comprehensive crane standard [www.osha.gov/FedReg_osha_pdf/FED20100809.pdf]

OSHA crane regulation text [www.osha.gov/cranes-derricks/index.html]

A press release for the crane standard [www.advancedsafetyhealth.com/blog/index.php/category/cranes]

OSHA PPE publications:

www.osha.gov/Publications/osha3151.html
www.osha.gov/OshDoc/data_General_Facts/ppe-factsheet.pdf
www.osha.gov/OshDoc/data_Hurricane_Facts/construction_ppe.pdf

Test Questions

1. What is the goal of PtD?

2. Give two examples of industries that have incorporated PtD into the corporate culture.

3. Name one practical benefit of PtD.

4. Give one ethical reason for PtD

5. Give an example of a hazard associated with an urban construction site.

6. What conditions might cause the sides of an excavation to cave in?

7. List three kinds of personal protective equipment (PPE).

8. Give three reasons why PPE is considered the solution of last resort.

9. How is PtD different from engineering controls?

10. Define "constructability."

11. Name the players who must communicate during the design phase.

12. When in the design process is the time to consider safety?

13. Why should you visit the OSHA Web site?

14. Name three construction hazards.

15. Where can you find tools to help you create safer designs?

Answers

1. The goal of PtD is to anticipate and eliminate hazards and risks at the design phase of a project/process and to make workplaces safer for workers.

2. Construction companies computer and communications corporations, design-build contractors, electrical power providers, engineering consulting firms, oil and gas industries, water utilities

3. Accidents on the job hurt employee morale, delay project completion, and cost money.

4. Preventable accidents should be prevented! Accidents ruin lives.

5. Examples include overhead power lines, existing infrastructure (gas, electric, and sewer), pedestrians, traffic flow.

6. A cave in can be caused by spring thaw, lack of shoring, cracked forms, recent precipitation, type of soil, and/or placement of heavy equipment

7. PPE includes items worn as a last line of defense against injury. OSHA-required PPE can include hardhats, steel-toed boots, safety glasses or safety goggles, gloves, earmuffs, full body suits, respiratory aids, face shields, and fall harnesses.

8. PPE is a solution of last resort because it
   a. requires the worker to wear it,
   b. may not fit because of limited size availability, and
   c. does not eliminate the hazard.

9. Engineering controls isolate the process or contain the hazard. PtD removes or reduces the hazard.

10. The term "constructability" implies an evaluation of a particular design in terms of cost, safety, duration, and quality. Can the design be built at a reasonable cost, within a reasonable amount of time, and result in an acceptable level of quality?

11. The entire design team must communicate, including the architect, structural engineer, civil engineer, HVAC engineer, trade representatives, and site planner.

12. Throughout!

13. OSHA regulations are updated annually. The Web site contains a summary of the latest hazard investigations. It also contains information about occupational diseases.

14. Hazards include falls, tripping hazards, falling objects, loud noises, and musculoskeletal injuries.

15. Agencies such as OSHA, NIOSH, and CHAIR can provide tools to help you create safer designs.